OFF THE CHARTS!
DATA INTERPRETATION

Crushing Standardized Test Math
for the GMAT®, GRE®, SAT®, PSAT/NMSQT®, and ACT®

Off the Charts! Data Interpretation, 1st edition

13-digit International Standard Book Number: 978-1-7320510-1-0

Copyright © 2019 by Adeptation

ALL RIGHTS RESERVED. No part of this work may be reproduced or used in any form or by any means – graphic, electronic, or mechanical, including photocopying, recording, taping, or web distribution – without the prior written permission of the publisher, Adeptation.

Notes:

GMAT® is a registered trademark of the *Graduate Management Admissions Council*, which neither sponsors nor endorses this product.

GRE® is a registered trademark of the *Educational Testing Service*, which neither sponsors nor is affiliated in any way with this product.

SAT® is a trademark registered by the *College Board*, which was not involved in the production of, and does not endorse, this product.

PSAT/NMSQT® is a registered trademark of the *College Board* and the *National Merit Scholarship Corporation*, which are not affiliated with, and do not endorse, this product.

ACT® is a registered trademark of *ACT, Inc.*, which neither sponsors nor endorses this product.

Table of Contents

Chapter 1 Welcome 1-1
Chapter 2 Data Interpretation: Tables, Charts & Graphs 2-4
    Best Practices for Creating Good Data Visualizations 2-4
    General Approach for Intrepreting Data Visualizations 2-10
    Types of Data Visualizations 2-11
    Describing Trends in Data 2-12
        Linear Relationships 2-12
        Exponential Relationships 2-13
        Correlations between Variables 2-15
    Types of Questions about Data Visualizations 2-16
    Common Wrong Answer Types for Data Interpretation 2-17
Chapter 3 Tables 3-19
    Concept Review 3-19
    Practice Sets 3-21
        Table 1: Labor Hours & Production by Factory 3-21
        Table 1: Interpreting the Data 3-23
        Table 1: Answers & Explanations 3-24
        Table 2: Plant Growth 3-30
        Table 2: Interpreting the Data 3-32
        Table 2: Answers & Explanations 3-34
Chapter 4 Pie Charts 4-37
    Concept Review 4-37
    Practice Sets 4-40
        Pie Chart 1A/1B: Net Revenue by Product Category 4-40
        Pie Chart 1A/1B: Interpreting the Data 4-42
        Pie Chart 1A/1B: Answers & Explanations 4-44
        Pie Chart 2: Patient Referral Sources 4-50
        Pie Chart 2: Interpreting the Data 4-51
        Pie Chart 2: Answers & Explanations 4-52
Chapter 5 Line Charts 5-54
    Concept Review 5-54
    Practice Sets 5-55
        Line Chart 1: Student Sick Days 5-55
        Line Chart 1: Interpreting the Data 5-56
        Line Chart 1: Answers & Explanations 5-58

Off the Charts! Data Interpretation

    Line Chart 2: Company Financial Performance 5-60
    Line Chart 2: Interpreting the Data 5-62
    Line Chart 2: Answers & Explanations 5-64

Chapter 6 Bar & Column Charts 6-69

  Concept Review 6-69

  Practice Sets 6-70

    Column Chart 1: Musical Instruments Played 6-70

    Column Chart 1: Interpreting the Data 6-71

    Column Chart 1: Answers & Explanations 6-73

    Column Chart 2: New Vehicle Sales at a Dealer 6-76

    Column Chart 2: Interpreting the Data 6-77

    Column Chart 2: Answers & Explanations 6-79

    Bar Chart 1: Applications by Metro Area 6-82

    Bar Chart 1: Interpreting the Data 6-83

    Bar Chart 1: Answers & Explanations 6-84

Chapter 7 Frequency Tables & Frequency Distributions 7-86

  Concept Review 7-86

    Version 1T: Frequency Table, Relative Frequency, Absolute Numbers 7-90

    Version 1D: Frequency Distribution, Relative Frequency, Absolute Numbers .. 7-91

    Version 2T: Frequency Table, Cumulative Frequency, Absolute Numbers 7-92

    Version 2D: Frequency Distribution, Cumulative Frequency, Absolute Numbers .. 7-93

    Version 3T: Frequency Table, Relative Frequency, Relative Numbers 7-94

    Version 3D: Frequency Distribution, Relative Frequency, Relative Numbers ... 7-95

    Version 4T: Frequency Table, Cumulative Frequency, Relative Numbers 7-96

    Version 4D: Frequency Distribution, Cumulative Frequency, Relative Numbers ... 7-97

    Version 5: Integrated Frequency Distribution for Side-by-Side Comparison 7-98

  Practice Sets 7-99

    Frequency Distribution 1: Number of Relatives 7-99

    Frequency Distribution 1: Interpreting the Data 7-101

    Frequency Distribution 1: Answers & Explanations 7-102

    Frequency Table 2: University Credit Hours 7-106

    Frequency Table 2: Interpreting the Data 7-107

    Frequency Table 2: Answers & Explanations 7-108

Chapter 8 Histograms 8-110

  Concept Review 8-110

  Practice Sets 8-114

    Histogram 1: First Smartphone 8-114

    Histogram 1: Interpreting the Data 8-116

    Histogram 1: Answers & Explanations 8-117
    Histogram 2: Freshman GPAs 8-122
    Histogram 2: Interpreting the Data 8-124
    Histogram 2: Answers & Explanations 8-125

Chapter 9 Box-and-Whisker Plots 9-129

Concept Review 9-129

Practice Sets 9-134

    Box-and-Whisker Plot 1: Language Development 9-134
    Box-and-Whisker Plot 1: Interpreting the Data 9-135
    Box-and-Whisker Plot 1: Answers & Explanations 9-136
    Box Plot 2: Athlete Stature 9-139
    Box Plot 2: Interpreting the Data 9-140
    Box Plot 2: Answers & Explanations 9-142

Chapter 10 Scatterplots 10-145

Concept Review 10-145

Practice Sets 10-150

    Scatterplot 1: Studying and Final Exam Scores 10-150
    Scatterplot 1: Interpreting the Data 10-151
    Scatterplot 1: Answers & Explanations 10-153
    Scatterplot 2: Urban Commute Times 10-156
    Scatterplot 2: Interpreting the Data 10-158
    Scatterplot 2: Answers & Explanations 10-160

Chapter 11 Paired & Mixed Data Visualizations 11-166

Concept Review 11-166

Practice Sets 11-171

    Paired Charts 1: Household Spending Data 11-171
    Paired Charts 1: Interpreting the Data 11-173
    Paired Charts 1: Answers & Explanations 11-175
    Paired Charts 2: Boutique Sales 11-178
    Paired Charts 2: Interpreting the Data 11-180
    Paired Charts 2: Answers & Explanations 11-182
    Mixed Charts 1: Student Reading Proficiency 11-184
    Mixed Charts 1: Interpreting the Data 11-186
    Mixed Charts 1: Answers & Explanations 11-188

Chapter 12 Refresher Content & Strategies 12-191

Translating Words to Math 12-191

    Math Clue Words and What They Mean 12-192

Ratios & Proportions 12-193

Off the Charts! Data Interpretation

    Translating Words to Math in Ratio Problems 12-194

    Part-to-Part vs. Part-to-total Ratios 12-195

  Percentages: Percent Of and Percent Change 12-196

  Probability 12-198

    Thinking through Logic: Words First, Math Second 12-202

  Statistics & Weighted Averages 12-203

    Understanding the Measures of Centrality 12-204

    Weighted Averages 12-205

    Percentiles, Deciles, Quintiles, and Quartiles 12-206

Chapter 13 Closing 13-208

CHAPTER 1 WELCOME

Off the Charts! Data Interpretation serves as a valuable supplement to whichever test-specific study guide you prefer. This book will help you learn to systematically make sense of data visualizations, decode the associated word problems, set up the correct equations, organize your scratch work to "error-proof" yourself, and efficiently arrive at the right answer, every time.

In this book, you will learn a set of **structured-but-flexible approaches** which can be **combined in different ways** to solve even the hardest questions on your standardized tests.

You'll also find **strategies for word problems** and **refresher content** for the most common topics included in the word problems about data visualizations, such as **percentages**, **ratios**, **probability**, and **statistics** concepts.

> **You'll find the refresher content in Chapter 12.**
>
> For more in-depth preparation focused on Word Problems, including practice questions with follow-along explanations, **check out the author's other book**, *All Your Word Problems Solved: Crushing Standardized Test Math*.

Data Interpretation. These are among the hardest problems in the quantitative sections of the various standardized tests.

Why are these problems the ultimate challenge for your quantitative reasoning skills?
- There are <u>many types</u> of **data visualizations**, a broad term which includes tables, charts, and graphs.
    - Each type of data visualization can come in <u>many subtle variations</u>.
    - Numbers can be presented in several different ways.
- You need to correctly interpret the wording of questions about them, so you know **what information you must extract** from the data visualizations.
    - <u>Simple</u> data interpretation questions ask you to <u>extract</u> information.
    - <u>Complex</u> data interpretation questions ask you to <u>extract</u> and <u>transform</u> this information in some mathematical way.
- You may need to **translate the word problem** on the page into a math problem you can solve.

Off the Charts! Data Interpretation

- You need to be comfortable looking at data visualizations and confident in your ability to **correctly extract information** from them. This, in turn, is dependent upon your visual-spatial processing and reasoning skills.
- You need to then know how to input the information from the data visualization into some formula and then perform any required calculations to arrive at an answer.

In essence, data interpretation questions challenge students because to solve them, you need to integrate your skills related to interpreting word problems, interpreting data visualizations, setting up correct equations, and doing math, all while avoiding careless mistakes.

Most major publishers only <u>briefly</u> explain data interpretation problems and demonstrate <u>simple</u> examples in their study guides, then challenge students to make the leap to solving <u>very difficult</u> practice problems. Even students in test prep courses offered by these major publishers remain confused and lacking confidence about how to approach data interpretation problems.

How do I know this?

Because I have **extensive** teaching, tutoring, and academic coaching experience, developed in parallel with my corporate experience.

Because I also have **extensive** corporate and consulting experience, working with data and helping other professionals to <u>understand</u> and <u>determine what actions to take</u> based upon key performance indicators (KPIs) in their management reports and dashboards.

The author's teaching, tutoring, and academic coaching experience includes not only standardized test prep, but also a vast array of math-intensive courses at the college, graduate, and MBA levels.

In other words, **this author has seen a lot of people do a lot of quantitative subjects** over the years and has accumulated deep insight about **why some students struggle with understanding and retaining concepts**, **why good students make mistakes**, and how to help you **get the right answer on the first try**.

My teaching, tutoring, and academic coaching experience spans all these subjects:

- College Admissions Test Prep: PSAT/NMSQT, SAT, and ACT
- Graduate Admissions Test Prep: GMAT and GRE
- Math: Algebra 1, Algebra 2, Geometry, Trigonometry, Precalculus, Quantitative Reasoning
- Economics: AP, IB, College, Graduate & MBA Microeconomics and Macroeconomics, with algebra and with calculus
- Finance: College & MBA finance courses
- Accounting: College & MBA Financial Accounting, Managerial Accounting
- Statistics & Related Courses: College, Graduate & MBA Statistics, Market Research Methods, Research Methods for Psychology
- Business & Marketing: Decision Models, Consumer Behavior, Microsoft Excel

My corporate and consulting experience includes:

- Working as a management reporting analyst, consultant, and Chief of Staff
- Helping sales executives to understand and act upon their business unit's reporting metrics
- Creating custom analyses for C-suite executives and their teams
- Creating data visualizations for presentation to audiences of varying abilities to interpret them
- Extracting, analyzing, and creating data visualizations using spreadsheet, data analysis, and/or dashboard reporting software programs, as clients requested

What makes this book dramatically more useful than most other books out on the market?

- The **simplification** of approaches and formulas, plus intentional communication in ways which make things **sticky and memorable**, so you recall it on test day. Most authors of math books explain things in a very academic way, which can be both frustrating and not particularly memorable. I want you to gain an appreciation for data interpretation and understand how this skill relates to your professional life, so that you learn that **data interpretation problems are much more intuitive than they seem at first**. Occasionally you will find some parts of this book are repetitive; that's intentional to reinforce the connections you should be making across topics and/or approaches. Other parts are highly detailed, and feel dense; that's also intentional, so you can prepare yourself for the seemingly infinite variations and complexities a good test question can introduce into an otherwise familiar topic.

- The **depth of explanation**, not only illustrating how to arrive at the right answers, but also explaining why the wrong answers are wrong and identifying what mistakes you may have made to arrive at those wrong answers. Enough variations in the type of data visualization, type of interpretation questions and question wording to get you prepared for your test.

- **Written entirely by a single author**, with consistent strategies throughout the book, so you can easily refer to a topic previously covered.

- **Written with the needs of students with mild to moderate learning disabilities in mind**. Transforming math into verbal processes and supporting interpretation of multi-step problems using a **color-coding system** is a game changer. This book was written, in large part, to enable one of my favorite clients, and students like her, to learn data interpretation in a more intuitive way.

This book is dedicated to the client who inspired this book and to two key people:

1. My dad. It simply never occurred to him that his daughter shouldn't be great at math. Moreover, he has encouraged me to dream bigger and explore new opportunities. He landed the job of editor of this book because when I showed him a preview of my new book, he immediately pointed out at least a dozen times where I had forgotten to use the plural verb with the Latin word "data."

2. My good friend Amelia. She's brilliant and supportive. She was the editor of my first book and cheerleader while I wrote this one.

CHAPTER 2 DATA INTERPRETATION: TABLES, CHARTS & GRAPHS

Data interpretation is a skill which requires you to make sense of information presented in visual forms, such as tables, charts, and graphs. There are many types.

If you have experienced any of the following:
- A tendency to mix up left and right
- A tendency to mix up the X-axis (horizontal) and Y-axis (vertical) of graphs
- Difficulty reading printed maps
- Difficulty drawing straight lines

Then you may find data interpretation questions to be extra challenging. Consider trying the following techniques to help you work with data visualizations and accurately retrieve information from data visualizations:

- **Use a piece of blank paper to hide the irrelevant parts of the visual**, so you can more easily focus on the part which is relevant.
- **Use any available object to create a straight line** your eye can follow, from the dot on a chart to its axis, or across a row of a table, or down a column of a table. Objects which may be available include: a piece of scratch paper, a second pen or pencil, your library card or other card in your wallet, etc.
- **Use graph paper to organize your scratch work**, as you set up equations and place the information from the data visualization into its corresponding place in the equation.

If you have ever had an IEP (Individualized Education Plan) or suspect you may have a learning disability, consider reaching out to a specialist for additional ideas for practical accommodations.

BEST PRACTICES FOR CREATING GOOD DATA VISUALIZATIONS

To improve your ability to interpret data visualizations, it is useful to understand what it takes to create good, clear ones, so you will be more attuned to what can make others unclear. When you are a working professional, you will want to make data visualizations which are clear and easy to follow. Standardized tests, however, are <u>testing your ability to figure things out</u> for yourself when the data visualizations are not-so-easy to follow.

1. Identify and classify your variables.
2. Choose an appropriate visualization for the kind of data you have.
3. Include the requisite descriptive attributes needed for the type of data visualization you created.
4. Implement thoughtful visual design choices to facilitate readability and ease-of-comprehension.

Identify and classify your variables. Understand the types of variables so you can:
- Identify independent or dependent variables for science-related questions.
- Articulate how an experiment was designed or how the variables are related.
- Create your own graphs in a way which make sense to others.

Independent Variable	Dependent Variable
Independent variables are the ones that: • The researcher controls or manipulates intentionally (or uses as a classification) • Are typically integer values • Are in evenly spaced increments	Dependent variables are the ones that: • The researcher observes happen (an outcome, but not necessarily the result of a cause-and-effect) • Are often decimal values • Are not always in evenly spaced increments
Memory trick: independent, intentionally, integer, and increments all start with the same prefix "in-"	Memory trick: dependent and decimal start with the same prefix "de-"
For graphs in the XY-plane, plot on the X-axis	For graphs in the XY-plane, plot on the Y-axis
For tables, list the independent variable(s) in the left-hand column(s)	For tables, list the dependent variable(s) in the right-hand column(s)

Discrete Variable	Continuous Variable
Discrete variables are the ones which can only be integer values.	Continuous variables are ones which can be either integer or non-integer values (i.e., decimals or fractions).
Examples of discrete variables: • Number of items purchased • Number of units produced • Number of people at a party • Number of flight segments	Examples of continuous variables: • Household income • Height of people or objects • Business revenue • Number of miles flown
Discrete variables mean you can make and sell either 8 or 9 or 10 units of a good, but you cannot make and sell partial units... ...so 8.2 or 9 ½ or 10.333 units are not possible values for the number of units made and sold.	Continuous variables mean both integer and non-integer values are valid... ...so you can have a household income of exactly $65,000, but $64,897.23 and $65,204.11 are also possible values.

Off the Charts! Data Interpretation

Choose an appropriate visualization for the kind of data you have. You may be asked whether certain data are best presented in a table, pie chart, line graph, or other form. The choice depends on the <u>level of measurement</u> and the <u>main idea you need to convey</u>, e.g., how parts contribute to a whole or how variables change over time.

Level of Measurement	Examples	Appropriate Visualizations
Nominal Descriptive, categorical information, which has no directionality	Counting people with various eye colors Blue eyes are different, not better or worse, than brown eyes, so the data are not directional	Tables Pie charts Bar & column charts Frequency tables Pictographs
Ordinal Descriptive, categorical information, which has directionality, but the values are not equally spaced	Ratings or segmentation data, such as: Good/Better/Best Disagree/Neutral/Agree You can't quantify how much more or less one rating is than another	Tables Pie charts Bar & column charts Histograms Frequency distributions Pictographs
Interval Numeric information, which has directionality and the values are equally spaced, but one value is not a multiple of another value (and the zero is not a true zero)	Temperature in Fahrenheit or Celsius Each degree is the same amount, but you cannot conclude that 55°F is 10% warmer than 50°F Shoe sizes US shoe sizes increase by 0.5 consistently, but you cannot conclude that size 10 is twice as large as size 5	Tables Pie charts Bar & column charts Histograms Frequency distributions Pictographs Line charts
Ratio Numeric information, which has directionality, the values are equally spaced, and each value is a multiple of another value (the zero is meaningful and absolute)	Height Length Each inch or centimeter is the same amount, and there is an absolute zero, so you can conclude that 70 inches is twice as long as 35 inches.	Tables Pie charts Bar & column charts Histograms Frequency distributions Pictographs Line charts & scatterplots Box-and-whisker plots

Include the requisite descriptive attributes needed for the type of data visualization you created. *Remember, when you are a working professional, you need to include these attributes to make your data visualizations easy for your coworkers and clients to interpret.* <u>*Standardized tests, however, will often omit these*</u> *or make them less clear than they need to be. Always ask yourself* <u>*if these elements could be clearer*</u>*.*

Required Attributes	Best Practices
Title	The title of any chart or graph should answer four questions: (1) Who, (2) What, (3) When, (4) Where Examples: • Incomplete: Clothing exports from China • Better: Clothing exports from China by destination country, for 2015 • Incomplete: Cost of textbooks at College A • Better: Average annual student spending on textbooks at College A, by student major, in 2010
Labels, headings, and/or axis titles with units of measurement	The labels, headings, and/or axis titles should be clear about: • What is being measured • In what <u>units of measurement</u>, e.g., meters vs. kilometers. Do "sales" represent dollars of revenue or number of items sold? • Whether <u>place value</u> has been manipulated, e.g., "in 000s of dollars" or "revenue in millions of Euros"
Legend (also called a key)	A legend is needed to indicate: • Which line is which, on a line graph with multiple lines • What category each sector of a pie chart represents • What category each column or bar represents, on clustered and/or stacked column and bar charts • What the bubble size represents on a bubble chart • What each icon or image represents on a pictograph
Footnotes	Footnotes can be used for any of these reasons: • To cite source(s) of the data • To call out any data points which require an explanation or clarification, such as a change in data collection or measurement methods (such as surveys completed by phone vs. by internet in different years) or adjustments made to ensure comparability

Off the Charts! Data Interpretation

Implement thoughtful visual design choices to facilitate readability and ease-of-comprehension.

Visual Design Choices	Best Practices
Number of categories	If you have more than 2-7 categories, it may be best to collapse the smaller categories into an "all other" category. Doing so will allow you to: • Save space on the page or screen. • Make the text and numbers describing the larger or more meaningful segments large enough to read.

Use of color	Determine whether you can use color or need to use a black-and-white texture or grayscale. Usually the latter options are used when data visualizations must be printed at minimum cost.
	
	If you are using color, consider:
	
	• Individuals may be partially or totally colorblind. Avoid red/green specifically (the most common partial colorblindness) and use blue/orange instead.
	
	• Whether there is enough contrast among the various colors, as well as between each color and the corresponding text to be readable.
	
	• Alternating both the color itself and light/dark to better visually distinguish parts of charts which are touching each other. For example, here are some combinations to avoid and one which would be better:
	
	    o Avoid: Purple – Blue – Green – Red. *These are the same level of light/dark.*
	
	    o Avoid: Dark Blue – Medium Blue – Light Blue. *These are the shades of the same color.*
	
	    o Better: Blue – Light Peach – Violet – Light Green. *These alternate both light/dark and color.*
	
	• Whether certain colors have common, default, or dominant interpretations which will override viewer's interpretation of your data visualization and legend.
	
	    o Avoid: Pure Red. *This means "stop" in traffic signals, "roadblock" in project management, and "losing money" in finance.*
	
	    o Better: Maroon or Pink. *These are safer to use because they do not have the same default interpretations that Pure Red does.*

Visual Design Choices	Best Practices

Readability of text and numbers	For many types of data visualizations, you can add data labels. These labels may contain one or more of the following items:
	
	- Category name
	- Absolute number (value). *Examples include 15 or 330,000 or 22 million.*
	- Relative number (percentage or fraction). *Examples include 20% or one-quarter of the prior amount.*
	
	You can choose whether to label:
	
	- All the data points
	- None of the data points
	- Only selected data points. If this is the case, the few data points with labels are significant in some way.
	
	Generally, putting all the information in the labels on all of the data points *seems* like a great idea, but this will result in a very cluttered data visualization.
	
	You may have seen graphs which **truncate numbers** into thousands, millions, or billions, which is done for two reasons:
	
	- <u>Magnitude matters more than precision</u>, especially for the intended audience. In other words, executives will care whether revenue is $1.5 million versus $1.2 million, but they will be less concerned whether it is precisely $1,561,288 or $1,572,345.
	- Rounding the numbers and trimming the extra digits enable the creator of the data visualization to <u>make the font size larger</u> and thus more readable.
	- You should inspect these places to determine if numbers were rounded to an order of magnitude:
	    - In the story above the data visualization
	    - Under the title of the chart or graph
	    - After the X-axis and/or Y-axis labels
	    - In a footnote below the data visualization
	
	When numbers have been rounded to an order of magnitude, take extra care with handling place value. Notation varies:
	
	- Billions: "in $B" or "in billions of units"
	- Millions: "in $M" or "in 000,000 units"
	- Thousands: "in $K" or "units (in 000s)"
	
	Units matter. *Always ask if something like "sales" or "product returns" are measured in dollars or numbers of items.*

GENERAL APPROACH FOR INTREPRETING DATA VISUALIZATIONS

You can follow this general approach to ensure you accurately interpret information presented visually.

1. **Read the story** or other text provided on the page which describes the context - the purpose of the data and the process of gathering that data.
2. **Read the title** of the chart or graph.
3. **Read the labels** and **note the units of measurement for each variable**.
   - For tables: Read the column headings and the row labels
   - For graphs in the XY-plane: Read the axis labels and determine the scale of the major & minor gridlines.
     i. The **major gridlines** are typically labeled on the graph in consistent increments.
     ii. The **minor gridlines** are often not labeled, but you can quickly figure out the size of these increments. If you start at one major gridline, count how many minor gridlines you go to the next major gridline. Then, divide the size of each major increment by the number of minor increments to determine the size of each minor increment.
     iii. When gridlines are present, you may find it **helpful to count from gridline to gridline crossed** (or box to box if you have both horizontal and vertical gridlines) by a bar or column. *This is less effortful than trying to read each upper and lower numeric value and perform subtraction multiple times.*
4. **Read the legend** (also called a **key**), if one is provided.
5. **Get a sense of the data** without dwelling on the details. Focus on the big picture.
   - Verify your understanding by trying to interpret the value of a *specific* data point, slice of a pie chart, or similar. If you can articulate in words, you probably understand it correctly. You may need to try articulating the meaning of the chosen data point a couple different ways until you land on the correct interpretation.
   - Look for **trends** or relationships in the data.
   - Look for **extremes**, such as **maximums**, **minimums**, or **outliers**.
   - Look for **changes in the data**, also called **inflection points**, where the graph might bend, change direction, or simply flatten out.
6. **Read any footnotes and/or explanations** of items marked with an **asterisk***.
7. If the data visualization violated the best practices for proper data presentation (discussed in the previous section), such as using a vague title or labels, then try to clarify these for yourself.

Sometimes, you will see two or more visual representations of data paired together. For these, add these steps:

8. **Identify how the two data visualizations are related**. Consider what is the same and what is different between paired data visualizations.
   - Are the **variables** the same or different?
   - Are the **categories of information** the same or different?
   - Are the **time periods** the same or different?
   - Are the **subjects of the research study** (locations, populations, etc.) the same or different?
   - Is the second visualization a **comparison to the first visualization**, or does it convey more **detail about a subset of the first visualization**?

9. **Note whether the "size of the pie" differs**. If the same data were gathered in different time periods or for different populations, note whether the total changed from one time period or population to the next one.

If you practice and consistently follow the steps listed to orient yourself toward and make sense of the information presented in various data visualizations, you will **start to develop some intuition about what questions you may be asked about the data**.

TYPES OF DATA VISUALIZATIONS

Visual representations of data include the basic forms listed in the table below. Variants of these basic forms exist to convey extra information which could not otherwise be presented in the basic form.

More common / general / familiar	Less common / specialty / less familiar
Tables	Box-and-whisker plots
Pie charts	Box plots
Line charts	*Bubble charts**
Bar & column charts	*Radar charts**
Frequency tables & distributions	*Waterfall charts**
Histograms	*Sunburst charts**
Scatterplots (also called XY-scatterplots)	*Pictographs**
*Organizational charts**	*Infographics**
*Flow charts**	*Items in italics are beyond the scope of this book.

Off the Charts! Data Interpretation

DESCRIBING TRENDS IN DATA

If you are looking at a graph in the XY-plane, try to identify what type of relationship is represented:

- **Linear relationship** – the graph is a straight (or mostly straight) line
- **Exponential relationship** – the graph is curved
- **No relationship** – the graph does not have a consistent pattern
- **Other relationship** – the graph might go up, then flatten out (increase then plateau) or the graph might go down, then level off (decrease then plateau). Whenever a graph is not perfectly linear, look for **critical points**, such as a **minimum**, **maximum**, or **inflection point**.

An **inflection point** (where the graph "bends" or changes direction) indicates something meaningful has changed in the relationship between the two variables, so you should anticipate a question about <u>when</u> or <u>why</u> the relationship may have changed.

LINEAR RELATIONSHIPS

With line charts and scatterplots with linear trendlines, there are two primary trends with multiple, synonymous ways to describe these two types of relationships.

Visual: Upward-sloping line	Visual: Downward-sloping line
Linear Direct (upward-sloping line)	Linear Inverse (downward-sloping line)
Y <u>varies directly</u> with X	Y <u>varies inversely</u> with X
As X increases, Y increases or As X decreases, Y decreases	As X increases, Y decreases or As X decreases, Y increases
X & Y move in the <u>same</u> direction	X & Y move in <u>opposite</u> directions
A <u>positive relationship</u> between X & Y	A <u>negative relationship</u> between X & Y

EXPONENTIAL RELATIONSHIPS

You can generally recognize that a graph expresses an exponential relationship when the **shape of the graph is curved**.

These are the two main categories:
- **Exponential Growth** – the graph is <u>increasing</u> from left to right
- **Exponential Decay** – the graph is <u>decreasing</u> from left to right

These each have two variations:
- **At an increasing rate** – the slope of the line gets <u>steeper</u> from left to right
- **At a decreasing rate** – the slope of the line gets <u>less steep</u> from left to right

Combined, these form the four primary exponential growth or decay trends which have multiple, synonymous ways to describe these four types of relationships.

Off the Charts! Data Interpretation

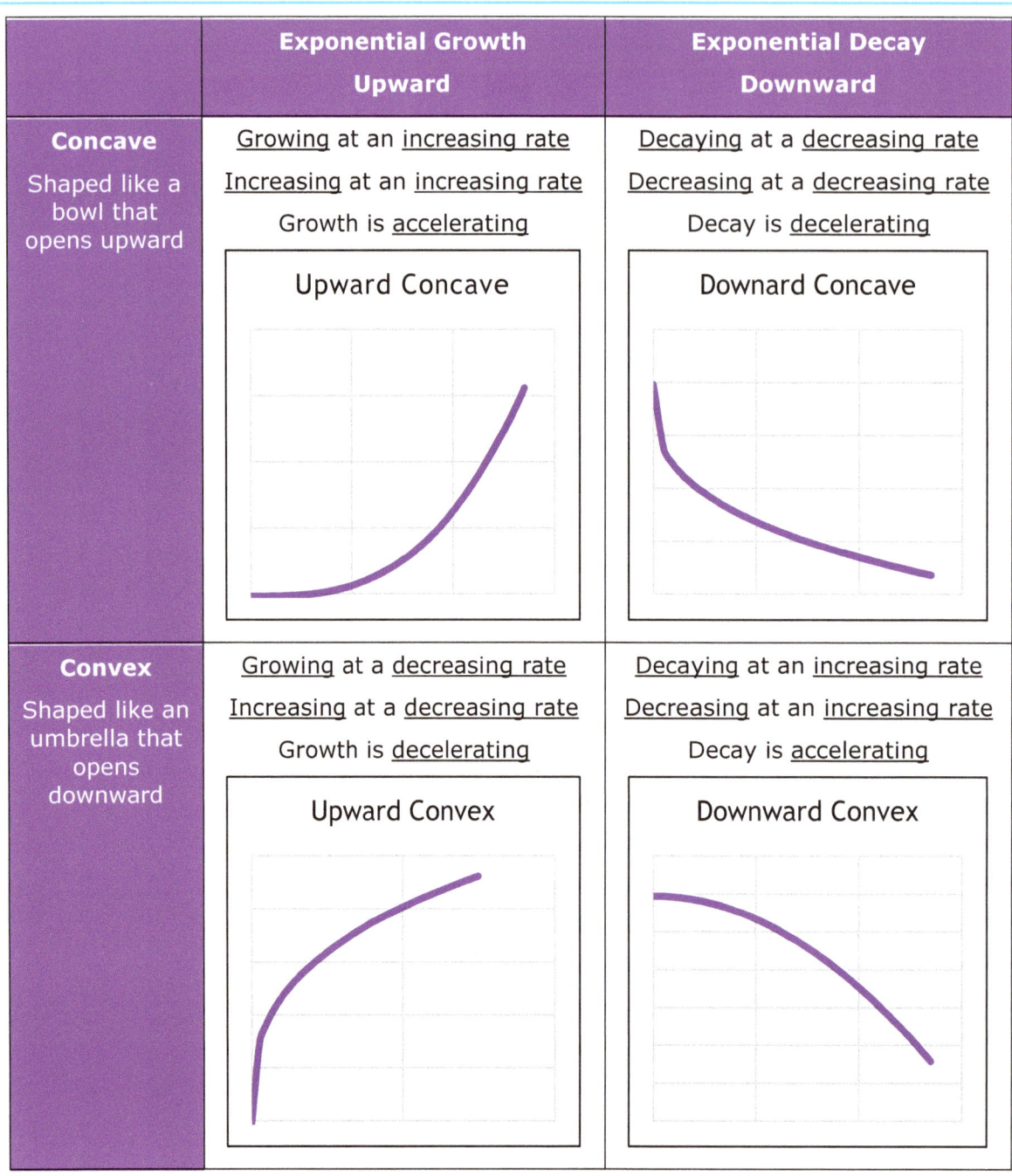

CORRELATIONS BETWEEN VARIABLES

For **scatterplots with trendlines**, you use an additional descriptor to indicate the strength of the relationship. You identify **strong** and **weak** correlations by examining how closely or loosely the various points fit to the trendline.

Strong Correlation	Weak Correlation
Strong Linear	Weak Linear
Strong Exponential	Weak Exponential
The individual (X,Y) plots closely follow the trendline.	The individual (X,Y) plots loosely follow the trendline.
Y is strongly correlated with X	Y is weakly correlated with X
Changes in the Y variable are mostly explained by changes in the X variable	Changes in the Y variable are somewhat explained by changes in the X variable
Most of the variation in the Y variable can be explained by variation in the X variable.	Some of the variation in the Y variable can be explained by variation in the X variable.

Off the Charts! Data Interpretation

TYPES OF QUESTIONS ABOUT DATA VISUALIZATIONS

Core Type	Variations
Part-to-part relative comparisons	These questions can be in the form of: • Ratios, of one value to another • Percent More or Percent Less • Percent Change
Part-to-total (part-to-whole) relative comparisons	You may be asked to compare the value of one part or category of the data to either a subtotal or the total population. These questions can be in the form of: • Ratios • Percent Of
Combining values	You may be asked to add two or more values from various parts or categories to calculate a subtotal or the grand total.
Absolute comparisons about two values	Value A is how much more / how much less than Value B

Statistical interpretations	• Mean (simple average), expected value, or weighted average
	    ○ Of <u>all</u> the data
	    ○ Of a <u>subset</u> of the data (e.g., the average for the 4th quintile vs. the overall average)
	• Median (middle value)
	• Mode (most common value)
	• Range
	    ○ Of <u>all</u> the data
	    ○ Of a <u>subset</u> of the data (e.g., the range of the 1st quartile vs. the overall range)
	• Maximum value
	• Minimum value

Probability of a certain outcome	• From a histogram • From a frequency distribution • From a normal distribution
Predictions about new data points, based on trends	• Extrapolate (go beyond the data set) • Interpolate (estimate between two data points)

When you approach standardized test questions about data visualizations, you'll want to use the following process:

- Read the question carefully, interpreting one phrase at a time, using the **Words to Math approach**.
- Structure your approach, using the **Words First**, **Math Second approach**.
- Carefully retrieve the necessary information from the data visualization(s) and place the values underneath the corresponding words to keep yourself organized.
- Solve for the answer as efficiently as possible.

Refer to Chapter 12, which provides general strategies for word problems, including those for:

- **Words to Math approach**
- **Words First**, **Math Second approach**

Chapter 12 also contains specific tips for common types of questions about data visualizations, such as:

- **Percent Of** and **Percent Change**
- **Ratios**
- **Statistics**

For more detailed information about working with Word Problems, **check out the author's other book**, *All Your Word Problems Solved: Crushing Standardized Test Math*

COMMON WRONG ANSWER TYPES FOR DATA INTERPRETATION

The nature of standardized tests implies that the wrong answer choices are standardized, and thus fall into patterns. People who create questions for these tests know that for any given type of question, there are certain patterns of mistakes that a large number of students make. Most of the available-but-wrong answer choices are thus the **result of errors in logic**, not mistakes in mental calculations. The logical mistakes students might make are far more predictable than the computational mistakes they might make.

The **logical errors** are typically related to one of three things:

- **Interpretation of the question's wording**. Some examples include:
    - **Misreading the question**

- o **Mixing up related-but-different terms**, such as Profit vs. Revenue, or Mean vs. Median, and as a result, *solving for the right answer to the wrong question.*
- o **Putting the wrong value in the denominator of a Percent Change question**
- o **Mixing up or not paying attention to what follows "of the..."** for percent change, ratios, or other questions.
- o **Mixing up whether you're solving for one part, a different part, or the whole**
- **Interpretation of the data visualizations**. Some examples include:
  - o **Reading data from the wrong axis**
    - This issue can happen more frequently when the values on both the X-axis and Y-axis are similar numbers.
    - On a double Y-axis chart, reading from the wrong axis. Watch out for graphs which have a second Y-axis on the right-hand side!
  - o **Mixing up or not paying attention to the legend**
    - On a dual-line or multiple-line graph, choosing a value from the wrong line
    - On a pie chart, choosing a value from the wrong sector (slice)
  - o **Not noticing when the graph does not start at the origin**
  - o **Not noticing when the "size of the pie" changed, when this would affect your approach and calculation of an answer**
  - o **Not combining information from multiple data visualizations correctly.** This issue is more common on the paired data visualizations.
- **Interpretation of numbers rounded to an order of magnitude.**
  - o Neither, one, or both variables may be rounded to orders of magnitude such as thousands, millions, or billions.
  - o Be aware that the different variables in a single data visualization or paired data visualizations may or may not be rounded to the same order of magnitude.
  - o For example, units of production might be provided in <u>thousands of units</u> and the revenue might be provided in <u>millions of dollars</u>. Failing to notice the different orders of magnitude could lead to calculation mistakes, if you are asked to calculate revenue per unit or revenue per thousand units.
  - o If you notice that variables are presented in different orders of magnitude, you should anticipate at least one question will test your ability to correctly handle **place value**.

CHAPTER 3 TABLES

CONCEPT REVIEW

A **table** is one of the most flexible types of data visualizations, because it can accommodate all four levels of measurement (nominal, ordinal, interval, and ratio) in a single format, while allowing you to quickly **compare** each group on different attributes.

When the goal of your data visualization is to highlight **trends**, however, tables are less useful than are line graphs, bar charts, and column charts.

There are many variations of tables which you might encounter. A **table** may display one or two variables. More **complex tables** may also be grouped to show categories and sub-categories, with or without **subtotals** and/or **grand totals**. The information in tables can be **grouped by rows** or **grouped by columns**.

The table below shows the average age, split by gender and by country for five countries.

- The creator did not show an all-gender average. Because the all-gender average would logically need to be a <u>weighted average</u>* and you do not have the size of the populations of each gender, you cannot calculate an all-gender average from the information in this table.
    - *Individuals may have children with multiple people at different times. One person's first is not necessarily the partner's first child.*
    - *Additionally, the number of females and males having children is not necessarily the same.*
- The creator did not show an all-countries average. Again, this would need to be a weighted average and you do not have the information in this table that you need to calculate an all-countries average.

Average Age at Birth of First Child		
	Females	Males
Country A	26.4	29.2
Country B	23.1	23.7
Country C	21.4	22.4
Country D	28.1	34.3
Country E	22.7	26.1

Off the Charts! Data Interpretation

The table below displays categories and sub-categories. It is **grouped by rows**.

- The cell (box) with the category name is a **merged cell** which groups the three product lines and the total row for the product category.
    - Thus, you know that Breads, Pastries, and Bagels are the three product lines grouped within the category of Bakery Items.
    - You also know that Sandwiches, Salads, and Soups are the three product lines grouped within the category of Café Items.
- This table shows the **total for each category** as well as a **grand total**.
    - Because the subtotals are shown in the same column as the amounts for each product line, the creator has chosen to bold the subtotals as one way to reduce confusion.
    - Alternatively, the subtotals could be highlighted in color (such as a pale grey), shown in a larger text size, and/or presented in a different column.
    - The choice of which visual treatments to apply to the subtotals and grand total is context-dependent.
    - By convention, whenever numeric data are presented in tables, the numbers should be right-aligned and decimal-aligned, to enable easy addition.

Product Category	Product Line	Sales Revenue (in $K)
Bakery Items	Breads	$ 466.2
	Pastries	$ 128.0
	Bagels	$ 65.5
	Total Bakery	**$ 659.7**
Café Items	Sandwiches	$ 920.1
	Salads	$ 344.5
	Soups	$ 265.0
	Total Café	**$ 1529.6**
Grand total		**$ 2189.3**

PRACTICE SETS

TABLE 1: LABOR HOURS & PRODUCTION BY FACTORY

A consultant captured the following information about Company X's four factories, which she organized into a table.

2017 Labor Hours & Production by Factory, at Company X

	Factory A	Factory B	Factory C	Factory D
Employees	450	220	130	270
Average Hours per Employee per Week	35.0	42.0	40.0	36.0
Total # of Units Produced per Week	38,000	28,000	13,500	23,000

Question 1: *Refer to Table 1.* The ratio of the number of employees at Factory D to the number of employees at Factory A is approximately...?

 a) 5 to 3
 b) 2 to 1
 c) 7 to 2
 d) 5 to 4
 e) 3 to 5

Question 2: *Refer to Table 1.* The number of units of produced at Factory B is what percent greater than the number of units of produced at Factory D?

 a) 18%
 b) 22%
 c) 41%
 d) 70%

Question 3: *Refer to Table 1.* Which factory produces the greatest number of units per labor hour?

 a) Factory A
 b) Factory B
 c) Factory C
 d) Factory D

Off the Charts! Data Interpretation

Question 4: *Refer to Table 1.* The number of units of produced per week at Factory A is approximately what percent of the total units produced per week at Factories A and B?

   a) 27%
   b) 37%
   c) 42%
   d) 58%

Question 5: *Refer to Table 1.* Which of the following statements are true, based upon the information in Table 1? Choose all that apply.

   a) The average employee at Factory A worked fewer hours per week than the average employee at any of the other factories.
   b) The number of labor hours at Factory D was less than the number of labor hours at Factory B.
   c) Each employee of Factory C worked more hours than the employees at Factory A.

TABLE 1: INTERPRETING THE DATA

A consultant captured the following information about Company X's four factories, which she organized into a table.

2017 Labor Hours & Production by Factory, at Company X

	Factory A	Factory B	Factory C	Factory D
Employees	450	220	130	270
Average Hours per Employee per Week	35.0	42.0	40.0	36.0
Total # of Units Produced per Week	38,000	28,000	13,500	23,000

To approach this table, start by noting:

1. **Read the story** – from this, you learn that the data are about one company with four different factories.

2. **Read the title** – 2017 Labor Hours & Production by Factory, at Company X
   - When – 2017 (presumably, the full year)
   - What – labor hours & production
   - Who/Where – Company X, which has multiple factories

3. **Read the labels** and **note the units of measurement for each variable**.
   - The four factories are listed horizontally across the top of each column
   - The three types of data are listed vertically down the left-hand column.

4. **Read the legend** – not applicable

5. **Get a sense of the data** without dwelling on the details. Focus on the big picture.
   - Factory A has the most employees and produces the most units. *(This makes sense).*
   - Factory C has the fewest employees and produces the fewest units. *(This makes sense).*
   - Factory B has fewer employees but produces more units than Factory D. *(Perhaps because they work different numbers of hours?)* Asking yourself questions like this is a great way to anticipate questions which could be asked about the data presented.

6. **Read any footnotes and/or explanations** – not applicable

Off the Charts! Data Interpretation

TABLE 1: ANSWERS & EXPLANATIONS

Question 1: *Refer to Table 1. The ratio of the number of employees at Factory D to the number of employees at Factory A is approximately...?*

   a) 5 to 3
   b) 2 to 1
   c) 7 to 2
   d) 5 to 4
   e) 3 to 5

- Recognizing the type of question: **Part-to-part comparison**. Specifically, it's asking for the **ratio** of info about one factory (a part) to the same info about a different factory (a part).
- Solving the question: Use **Words to Math** and **Words First, Math Second**.
    - "The ratio of..." → set up a fraction
    - "the number of employees at Factory D" → this comes first, so it goes in the numerator of your fraction
    - "to the number of employees at Factory A" → this comes second, so it goes in the denominator of your fraction
    - Retrieve the necessary pieces of information from the table. Next to the ratio, place the values for each to the right. Then, simplify.
    - $\frac{\text{\# of Employees at Factory D}}{\text{\# of Employees at Factory A}} = \frac{270}{450} = \frac{3}{5} = 3 \text{ to } 5$
    - Choose E.
- **Behind each wrong answer choice is faulty logic**:
    - *Choice A is incorrect because it is the reciprocal of the correct answer. Be cautious with ratios so that you put the right part in the right place!*
    - *Choice B is incorrect because it is the right answer to the wrong question. Here, it's the correct ratio for the number of employees at Factory D to the number of employees at Factory C. If you chose this one, you may have either misread the question or misread the table.*
    - *Choice C is incorrect because it is the right answer to the wrong question. Here, it's the correct ratio for the number of employees at Factory A to the number of employees at Factory C. If you chose this one, you may have either misread the question or misread the table.*
    - *Choice D is incorrect because it is the right answer to the wrong question. Here, it's the correct ratio for the number of employees at Factory D to the number of employees at Factory B. If you chose this one, you may have either misread the question or misread the table.*

Question 2: *Refer to Table 1. The number of units of produced at Factory B is what percent greater than the number of units of produced at Factory D?*

   a) 18%
   b) 22%
   c) 41%
   d) 70%

- Recognizing the type of question: **Part-to-part comparison**. Specifically, it's asking for the **percent comparison**, so use the percent change formula.
- Solving the question: Use **Words to Math** and **Words First**, **Math Second**.
    - "The number of units produced at Factory B" → what you are focused on
    - "Is what percent greater" → use the percent change formula
    - "Than the number of units produced at Factory D" → the basis of comparison
    - $Percent\ Change = \dfrac{Focus - Basis\ of\ Comparison}{Basis\ of\ Comparison}$
    - Retrieve the necessary pieces of information from the table. Next to the ratio, place the values for each to the right. Then, simplify.
    - $Pct.\ Change = \dfrac{Units\ at\ Factory\ B - Units\ at\ Factory\ D}{Units\ at\ Factory\ D} = \dfrac{28000-23000}{23000} = \dfrac{5000}{23000}$
    - $Pct.\ Change = \mathbf{21.7\%}$
    - Because this value is positive, interpret it as a 21.7% increase or 21.7% more.
    - Choose B.
- **Behind each wrong answer choice is faulty logic**:
    - Choice A is incorrect because it is the result of putting the wrong number into the denominator. Be cautious with percent change questions so that you put the right part in the right place!
    - Choice C is incorrect because it is the right answer to the wrong question. Here, it's the correct percent change for the number of units produced at <u>Factory D</u> compared to the number of units produced at <u>Factory C</u>. If you chose this one, you may have either misread the question or misread the table. If the latter, you retrieved data from the right row but the wrong columns of the table.
    - Choice D is incorrect in two ways. It is the result of comparing Factory D and Factory C, but also putting the wrong number into the denominator.

Question 3: *Refer to Table 1.* Which factory produces the greatest number of units per labor hour?

    a) Factory A
    b) Factory B
    c) Factory C
    d) Factory D

- Recognizing the type of question: **Statistical interpretation**. Specifically, it's asking about the greatest (**maximum** number), but instead of asking for that number, it is asking you to identify which factory is associated with that number.
- Solving the question: Use **Words to Math** and **Words First, Math Second**.
    - "Which factory produces the greatest" → You will need to calculate the number of units per labor hour for each of the four factories, identify the maximum, and pick the right factory from the list.
    - "Number of units" → The last row of the table provides you the number of units produced per week.
    - "Per" → This word indicates you need a fraction or ratio. Set up your fraction. *Adjustment:* Because the phrase "number of units" came before the word "per", then you know that "number of units" goes in the numerator of your fraction.
    - "Labor hour" → This phrase comes after the word "per," so you know that labor hours will go in the denominator of your fraction.
        - There is no row in the table called "labor hours."
        - You have one row with the number of employees – but that's not labor hours. You have one row with the number of hours per week per employee – but that's not labor hours either.
        - You must, therefore, combine these two pieces of information.
        - Labor hours = # of employees * # of hours per employee per week
        - Substitute this information into the original expression.
    - $$\frac{Number\ of\ Units}{Labor\ Hour} = \frac{\#\ of\ Units\ per\ Week}{\#\ of\ Employees * \#\ of\ Hours\ per\ Employee\ per\ Week}$$
    - **Use this formula to calculate the number of units per labor hour for each of the factories.** Retrieve the necessary pieces of information from the table. Plug in the values and simplify.
        - $A = \dfrac{\#\ of\ Units\ per\ Week}{\#\ of\ Employees * \#\ of\ Hours\ per\ Employee\ per\ Week} = \dfrac{38000}{450*35} = 2.41$
        - $B = \dfrac{\#\ of\ Units\ per\ Week}{\#\ of\ Employees * \#\ of\ Hours\ per\ Employee\ per\ Week} = \dfrac{28000}{220*42} = 3.03$
        - $C = \dfrac{\#\ of\ Units\ per\ Week}{\#\ of\ Employees * \#\ of\ Hours\ per\ Employee\ per\ Week} = \dfrac{13500}{130*40} = 2.60$
        - $D = \dfrac{\#\ of\ Units\ per\ Week}{\#\ of\ Employees * \#\ of\ Hours\ per\ Employee\ per\ Week} = \dfrac{23000}{270*36} = 2.37$
    - Choose B.

Question 4: *Refer to Table 1. The number of units of produced per week at Factory A is approximately what percent of the total units produced per week at Factories A and B?*

    a) 27%
    b) 37%
    c) 42%
    d) 58%

- Recognizing the type of question: **Part-to-part comparison**. Specifically, it's asking for the **percent of** info about one factory (a part) to the same info about two factories combined (a subtotal).
- Solving the question: Use **Words to Math** and **Words First, Math Second**.
    - "The number of units produced at Factory A" → this comes <u>first</u>, so it goes in the <u>numerator</u> of your fraction
    - "is approximately what percent of..." → set up a fraction
    - "the total units produced at Factories A and B" → this comes <u>second</u>, so it goes in the <u>denominator</u> of your fraction
    - Retrieve the necessary pieces of information from the table. Next to the fraction, place the values for each to the right. Then, calculate the percentage.
    - $$\frac{\text{\# of Units at Factory A}}{\text{Total \# of Units at Factory A and B}} = \frac{\text{\# of Units at Factory A}}{\text{\# of Units at Factory A} + \text{\# of Units at Factory B}}$$
    - $$\frac{38000}{38000+28000} = \frac{38000}{66000} = 57.6\% \text{ or } 58\%$$
    - *Mental math & reasoning shortcut: You could cancel out the zeroes and simplify the fraction to 19/33, which is more than half – and only one answer choice is available that is more than 50%. Alternately, you could multiply both numerator and denominator by 3 to get this to 57/99 which will be a little more than 57%.*
    - Choose D.
- **Behind each wrong answer choice is faulty logic**:
    - *Choice A is incorrect because it is the result of making both mistakes listed in Choice B and Choice C – both the wrong number in the numerator, and the wrong number in the denominator.*
    - *Choice B is incorrect because it is the result of putting the wrong number into the denominator. Here, if you had misread the question or failed to read the words past "total number of...", you might have put the grand total of units produced at all four factories into the denominator.*
    - *Choice C is incorrect because it is the result of putting the wrong number into the numerator. Be cautious with percent change questions so that you put the right part in the right place!*

Question 5: *Refer to Table 1.* Which of the following statements are true, based upon the information in Table 1? Choose all that apply.

   a) The average employee at Factory A worked fewer hours per week than the average employee at any of the other factories.
   b) The number of labor hours at Factory D was less than the number of labor hours at Factory B.
   c) Each employee of Factory C worked more hours than the employees at Factory A.

- Recognizing the type of question: Each statement may require a different approach, so you'll need to use the same general process but different specific steps.
- Solving the question: Use **Words to Math** and **Words First**, **Math Second**.
- Evaluate Statement A:
    - "The average employee at Factory A" → This information is in the table. Read on to the next phrase, and you'll see that you need to find the average hours per employee at Factory A. That's 35 hours per week.
    - "worked fewer hours per week than..." → set up an inequality
    - "the average employee at any of the other factories" → This information is in the table. From the previous phrase, you'll see that you need to find the average hours per employee at each Factory B, C, and D. That's 42, 40, and 36 hours per week, respectively.
    - Evaluate the truthfulness of the statement.
        - Is 35 hours per week < 42, 40, and 36?
        - Yes, 35 is less than all of those numbers. **Statement A is true**.
- Evaluate Statement B:
    - "The number of labor hours at Factory D" → There is no row in the table called "labor hours."
        - You have one row with the number of employees – but that's not labor hours. You have one row with the number of hours per week per employee – but that's not labor hours either.
        - You must, therefore, combine these two pieces of information.
        - Labor hours = # of employees ∗ # of hours per employee per week
        - Labor hours at D = 270 ∗ 36 = **9720**
    - "was less than..." → set up an inequality
    - "the number of labor hours at Factory B" → There is no row in the table called "labor hours."
        - You have one row with the number of employees – but that's not labor hours. You have one row with the number of hours per week per employee – but that's not labor hours either.
        - You must, therefore, combine these two pieces of information.
        - Labor hours = # of employees ∗ # of hours per employee per week

- - - *Labor hours at B* = 220 ∗ 42 = **9240**
  - ○ Evaluate the truthfulness of the statement.
    - ▪ *Is Labor Hours at D < Labor Hours at B?*
    - ▪ Is 9720 hours < 9240?
    - ▪ No, 9720 hours is not less than 9240 hours. **Statement B is false**.
- Evaluate Statement C:
  - ○ "Each employee of Factory C" – This information is not in the table. You only have the <u>average</u> hours per employee per week. Some employees might have worked more hours, and others fewer. You can draw comparisons from one average to another average, but you cannot draw any conclusions about individual workers or all the workers.
  - ○ "worked more hours than"
  - ○ "the employees at Factory A"
  - ○ **Statement C is false**.
- **You must choose only statement A, but not B or C, to get the point for the question**.

TABLE 2: PLANT GROWTH

A researcher, working with a major university's agricultural extension program, ran an experiment to determine the impact of varying the amounts of fertilizer given on a few popular varieties of berry plants. Berries were picked as they ripened, and the mass of berries picked each day was accumulated over the 90-day experiment. At the end of the experiment, the height of each plant was measured.

Fertilizer (g per 10L of water)	Variety	Height (cm)	Berry Mass (kg)
5	B1	60.5	5.1
5	B2	49.8	3.5
5	B3	68.7	5.2
5	B4	62.4	6.1
10	B1	72.3	7.1
10	B2	63.7	4.8
10	B3	80.2	6.9
10	B4	77.6	9.1
20	B1	94.1	9.2
20	B2	87.7	7.8
20	B3	79.5	7.1
20	B4	106.3	14.9

Question 1: *Refer to Table 2.* Which of the following is closest to the average mass in kilograms of berries collected across all varieties of berry plants, among those which received 10 grams of fertilizer?

   a) 7.0
   b) 7.2
   c) 73.5
   d) 75.2

Question 2: *Refer to Table 2.* Based upon the information in the table, compare the following quantities:

Quantity A	Quantity B
The increase in plant height for variety B4, when the amount of fertilizer is increased from 10 grams to 20 grams	The increase in plant height for variety B2, when the amount of fertilizer is increased from 5 grams to 20 grams

a) Quantity A is greater
b) Quantity B is greater
c) Quantity A and Quantity B are equal
d) Cannot be determined

Off the Charts! Data Interpretation

TABLE 2: INTERPRETING THE DATA

A researcher, working with a major university's agricultural extension program, ran an experiment to determine the impact of varying the amounts of fertilizer given on a few popular varieties of berry plants. Berries were picked as they ripened, and the mass of berries picked each day was accumulated over the 90-day experiment. At the end of the experiment, the height of each plant was measured.

Fertilizer (g per 10L of water)	Variety	Height (cm)	Berry Mass (kg)
5	B1	60.5	5.1
5	B2	49.8	3.5
5	B3	68.7	5.2
5	B4	62.4	6.1
10	B1	72.3	7.1
10	B2	63.7	4.8
10	B3	80.2	6.9
10	B4	77.6	9.1
20	B1	94.1	9.2
20	B2	87.7	7.8
20	B3	79.5	7.1
20	B4	106.3	14.9

1. **Read the story** – the data are about plant growth
2. **Read the title** – not applicable
3. **Read the labels** and **note the units of measurement for each variable**.
    - Fertilizer is the left-most column, so this is an **independent variable**, and is measured in grams per 10L of water.
    - Variety is listed in the next column. Variety is categorical, so this is a second **independent variable**.
    - Height in centimeters. The heights listed have decimal values, so height is a **dependent variable**.
    - Berry Mass in kilograms. The masses listed have decimal values, so berry mass is a **dependent variable**.

4. **Read the legend** – not applicable
5. **Get a sense of the data**
    - In the fertilizer column, each amount repeats 4 times before the next larger amount is listed. In the variety column, the variety codes B1, B2, B3, and B4 are listed and then the list repeats.
        i. Taken together, interpret these columns as "each of the 3 amounts of fertilizer was applied to each of the 4 varieties of berry plant."
        ii. You should anticipate questions which ask you to compare either the <u>same amount</u> of fertilizer on <u>different varieties</u> of berry plant, or <u>different amounts</u> of fertilizer on the <u>same variety</u> of berry plant.
        iii. On paper-based tests, you may find it helpful to darken the line between the groups to better visually distinguish among them.
    - Generally, both height and berry mass increase when larger amounts of fertilizer are used.
6. **Read any footnotes and/or explanations** – not applicable

TABLE 2: ANSWERS & EXPLANATIONS

Question 1: *Refer to Table 2. Which of the following is closest to the average mass in kilograms of berries collected across all varieties of berry plants, among those which received 10 grams of fertilizer?*

- a) 7.0
- b) 7.2
- c) 73.5
- d) 75.2

- Recognizing the type of question: **Statistical interpretation**. Specifically, it's asking for the **average**.
- Solving the question: Use **Words to Math** and **Words First, Math Second**.
    - "Which of the following is closest to the average mass in kilograms of berries..." → Determine whether you need a simple average or weighted average. You will use the data in the last column.
    - "collected across all varieties of berry plants" → You will need to include in your calculation the mass from all four plant varieties. *The lead-in here should also hint that something is coming next, which will narrow down which values to include in your calculation.*
    - "among those which received 10 grams of fertilizer" → Only consider the subset of values which relate to 10 grams of fertilizer.
    - Write your expression using the **Words First, Math Second** approach.
    - $Average\ Mass\ of\ Berries = \dfrac{Sum\ of\ Values}{Number\ of\ Values}$
    - $Average\ Mass\ of\ Berries = \dfrac{7.1 + 4.8 + 6.9 + 9.1}{4} = 6.975$
    - Choose A.
- **Behind each wrong answer choice is faulty logic**:
    - Choice B is incorrect. It is the average of all 12 berry mass values. If you chose this answer, you may have missed the detail in the question wording which instructed you to narrow the set of values to just those which received 10 grams of fertilizer.
    - Choice C is incorrect. It is the average of the plant height values, not the berry mass values, for the correct subset of plants. If you chose this answer, you may have rushed and retrieved data from the wrong column of the table.
    - Choice D is incorrect. It is the result of making both mistakes from choices B and C – averaging all 12 values *and* averaging the values from the wrong column of the table.

Question 2: *Refer to Table 2.* Based upon the information in the table, compare the following quantities:

Quantity A

The increase in plant height for variety B4, when the amount of fertilizer is increased from 10 grams to 20 grams

Quantity B

The increase in plant height for variety B2, when the amount of fertilizer is increased from 5 grams to 20 grams

a) Quantity A is greater
b) Quantity B is greater
c) Quantity A and Quantity B are equal
d) Cannot be determined

- Recognizing the type of question: For questions which ask you to compare to quantities (typically found on the GRE®), you may need to translate one or two statements from words to math.
- Solving the question: Use **Words to Math** and **Words First, Math Second**.
- Evaluate Quantity A:
  - "The increase in plant height for variety B4" → You will need to subtract to find the change in the plant height for this variety. *Note: The values you must compare are not in cells that touch each other, so be careful when you retrieve the two values from the table.*
  - "when the amount of fertilizer is increased from 10 grams" → Find the plant height for variety B4 when it receives 10 grams of fertilizer. The keyword "from" indicates this is your original value.
  - "to 20 grams" → Find the plant height for variety B4 when it receives 20 grams of fertilizer. The keyword "to" indicates this is your new value.
  - $Increase\ in\ Height\ =\ Height\ of\ B4\ with\ 20g\ -\ Height\ of\ B4\ with\ 10g$
  - $Increase\ in\ Height\ =\ 106.3 - 77.6 = \mathbf{28.7\ cm}$
- Evaluate Quantity B:
  - "The increase in plant height for variety B2" → You will need to subtract to find the change in the plant height for this variety. *Note: The values you must compare are not in cells that touch each other, so be careful when you retrieve the two values from the table.*
  - "when the amount of fertilizer is increased from 5 grams" → Find the plant height for variety B4 when it receives 5 grams of fertilizer. The keyword "from" indicates this is your original value.
  - "to 20 grams" → Find the plant height for variety B4 when it receives 20 grams of fertilizer. The keyword "to" indicates this is your new value.

- - $Increase\ in\ Height\ =\ Height\ of\ B2\ with\ 20g\ -\ Height\ of\ B2\ with\ 5g$
  - $Increase\ in\ Height\ =\ 87.7\ -\ 49.8\ =\ \mathbf{37.9\ cm}$
- Compare your answer for Quantity A to the number in Quantity B.
- **Quantity B is greater**.
- **Choose B**.

The most common incorrect answer for this GRE-style quantity comparison question would be to choose A, that quantity A is greater. If you chose A, you may have rushed through reading the second statement and assumed that the two amounts of fertilizer being compared (20 grams vs. 5 grams) were instead the same as in the first statement (20 grams vs. 10 grams).

CHAPTER 4 PIE CHARTS

CONCEPT REVIEW

A **pie chart** is one of the more common types of data visualizations and is often used because it can show the **relative** quantities among categories, allowing you to quickly compare each group on a single attribute as well as understand the contribution of each part to the total.

A **circle graph** is another name for a **pie chart**. Throughout this book, the term **pie chart** will be used.

When the goal of your data visualization is to highlight **trends**, however, pie charts are less useful than are line graphs, bar charts, and column charts.

For the information captured in a **pie chart** to be meaningful, the various categories must be **mutually exclusive and collectively exhaustive**. In other words:

- All possible categories must be included; none are omitted.
- All items must belong to exactly one category (sometimes smaller categories are grouped together into one called "all other").
- The sum of the parts equals the value of the whole.

Refer to Chapters 7 and 12 for more information about mutually exclusive and collectively exhaustive.

Pie charts typically present numeric information in one of three ways:

- Showing each <u>value</u>.
- Showing each <u>percentage</u> of the total.
    - If there is a **total value** provided in either the **story** or the **chart title**, then you can multiply each percentage times the total value to get each corresponding category value.
    - If there is <u>no</u> total value provided in either the **story** or the **chart title**, then you can only discern the **relative size** of each category, but not the **absolute size** of each category.
- Showing both the **absolute numbers** and the **percentages**. (This may be common professionally, but this is uncommon on standardized tests – you might be given absolute numbers and asked to find either a percentage or ratio or vice-versa).

There are many variations of pie charts.

- Not every pie chart needs a **legend**. If the category labels are shown on the chart itself, then a legend is not needed.
- The **slices** of the pie chart may be touching, but one or more might be separated slightly from the rest. When a single slice of a pie chart is separated from the rest of the pie, the purpose of doing so is to highlight this specific category and its data.

Off the Charts! Data Interpretation

- A pie chart might have a hollow center. In this case, it may be called a **donut chart** or **ring chart**. Donut charts have an advantage over pie charts because these could present the relative mix of more than one variable, by using an inner ring and an outer ring.

The pie chart shown below provides both the **category names** and the **absolute number** of responses for each category next to their respective slices.

- When needed, **leader lines** are used to more clearly connect the text to its corresponding slice. *Refer to the diagram below. A leaderine is used to connect Water Skiing with the purple slice of the pie chart.*

- The relative size of each slice conveys the relative popularity of each of the four sports. Wakeboarding appears to be approximately one-fourth of the total, whereas rafting appears to be a little less than one-fourth of the total.

- To calculate exact percentages, you would need to find the total of the four values by adding them, because the total number of survey responses was not provided to you.

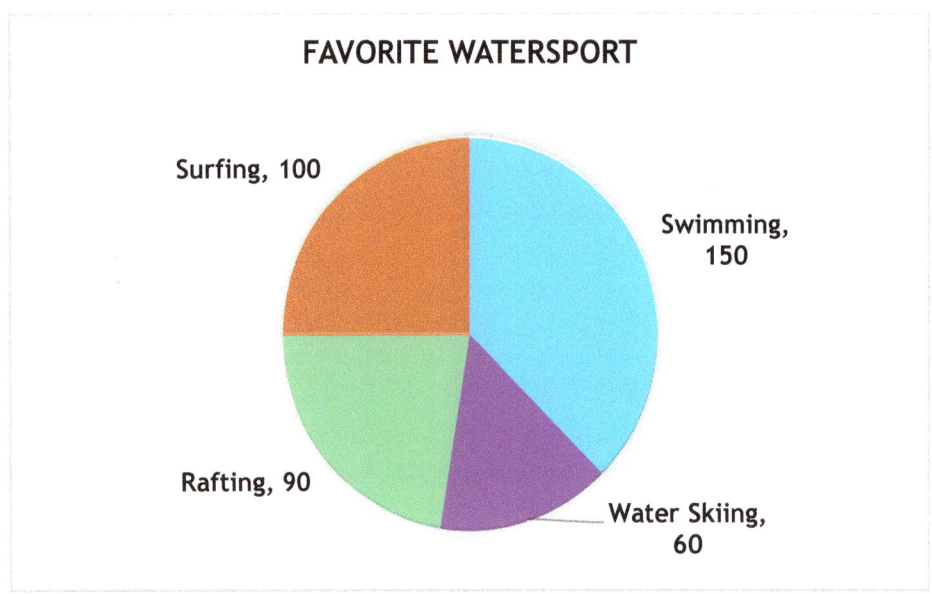

Example: For what percent of survey respondents is surfing their favorite watersport?

$$\frac{Surfing}{Surfing + Swimming + Water\ Skiing + Rafting} = \frac{100}{100 + 150 + 60 + 90} = \frac{100}{400} = 25\%$$

Off the Charts! Data Interpretation

The **donut chart** shown below provides both the **category names** and the **relative number** (percentage) of patients with each ailment next to their respective slices.

- The relative size of each slice conveys the relative incidence or frequency of each of the five ailments.
- The **legend** is shown below the **donut chart**.
- To calculate exact numbers, you would need to know the total number of patients and multiply the total by the five different percentages to find the number of patients with each type of ailment.
- For both pie charts and donut charts, the <u>total number</u> (in this case, of patients) may be provided:
    - In a paragraph above the chart
    - In a footnote below the chart
    - Below the chart title
    - Omitted

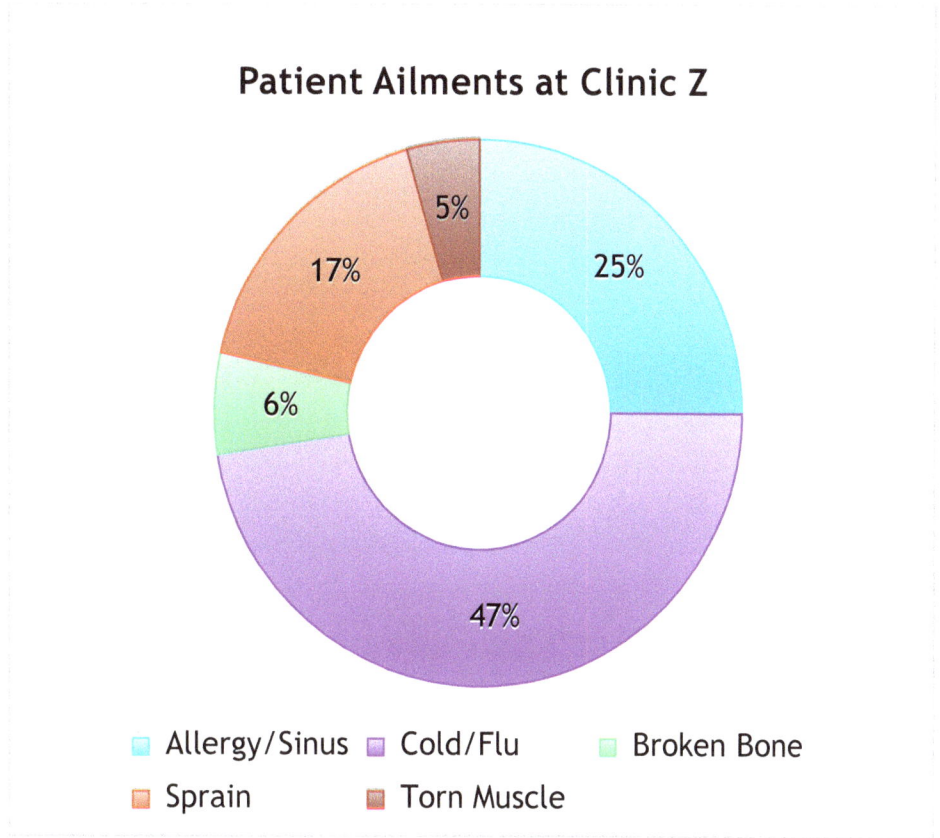

Example: if there were 500 patients, then there were how many patients with a sprain?

$$\# \text{ of Patients with Sprain} = \% \text{ of Patients with Sprain} * \text{Total \# of Patients}$$

$$\# \text{ of Patients with Sprain} = 17\% * 500 = \mathbf{85}$$

PRACTICE SETS

PIE CHART 1A/1B: NET REVENUE BY PRODUCT CATEGORY

The owner of Frank's Fine Furniture asked the store manager to prepare information about the store's revenue, showing the relative size of the seven categories of products sold in the store. The store manager was unsure what type of chart the owner liked to look at, so he prepared two different pie charts using the sales data for the past 12 months. *The same data was used to create two versions of pie charts.*

Pie Chart 1A

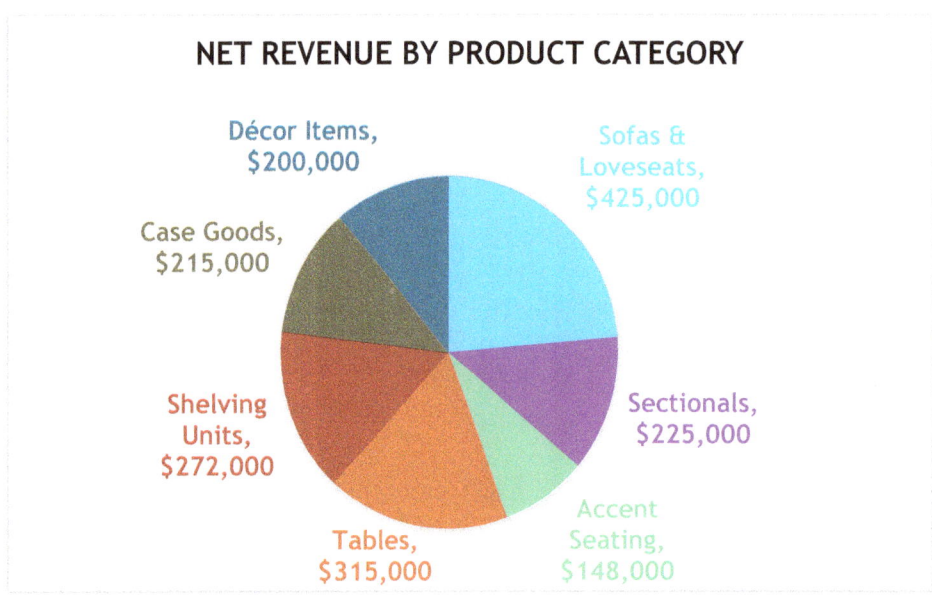

Pie Chart 1B

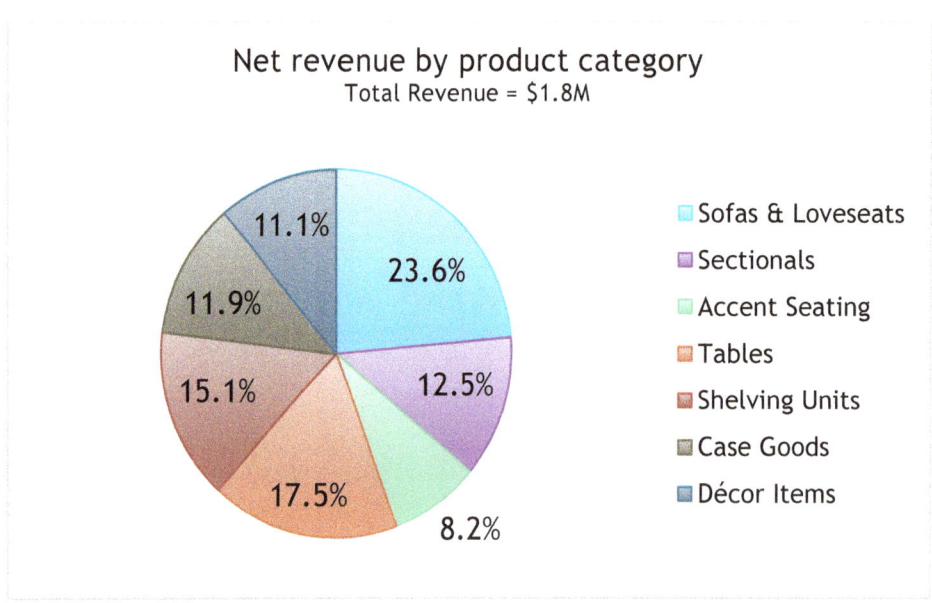

Question 1: *Refer to Pie Chart 1A.* The three highest-revenue categories generate approximately what percent of the total revenue at Frank's Fine Furniture?

　　a) 31.3%
　　b) 43.8%
　　c) 44.3%
　　d) 53.6%
　　e) 56.2%

Question 2: *Refer to Pie Chart 1B.* Approximately how much revenue did Frank's Fine Furniture generate from sales of Shelving Units and Case Goods?

　　a) $214,000
　　b) $272,000
　　c) $315,000
　　d) $486,000
　　e) $529,000

Question 3: *Refer to Pie Chart 1B.* Approximately how much was the median category revenue at Frank's Fine Furniture?

　　a) $214,000
　　b) $225,000
　　c) $257,000
　　d) $315,000
　　e) $425,000

Question 4: *Refer to Pie Chart 1A.* Which of the following statements are true, based upon the information in the pie chart? Choose all that apply.

　　a) The range of category revenues is $277,000.
　　b) The revenue from the Tables category is more than 150% of the revenue from Décor Items.
　　c) The revenue from the Sofas & Loveseats category is more than 150% more than the revenue from Décor Items.

Off the Charts! Data Interpretation

PIE CHART 1A/1B: INTERPRETING THE DATA

The owner of Frank's Fine Furniture asked the store manager to prepare information about the store's revenue, showing the relative size of the seven categories of products sold in the store. The store manager was unsure what type of chart the owner liked to look at, so he prepared two different pie charts using the sales data for the past 12 months. *The same data was used to create two versions of pie charts.*

Pie Chart 1A

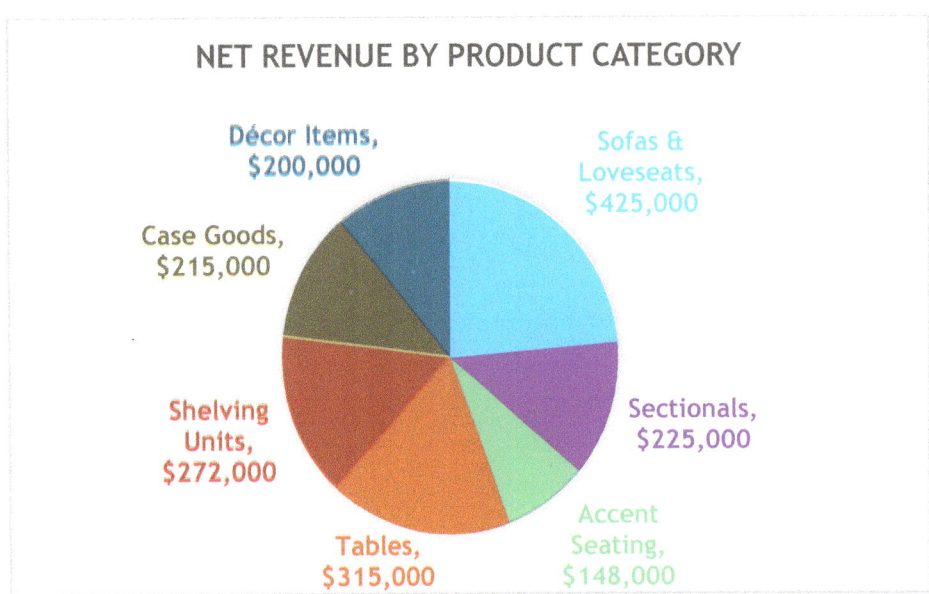

Pie Chart 1B

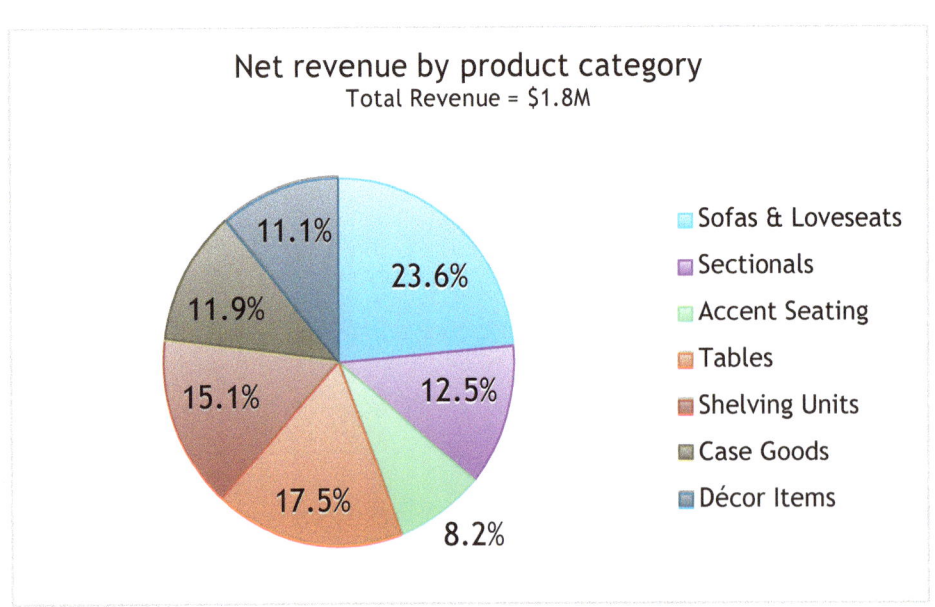

1. **Read the story** – from this, you learn that the data are about one furniture store with seven different product categories.

2. **Read the title** – Net revenue by product category

    - When – not shown in the title, but the **story** tells you this is for the past 12 months
    - What – net revenue
    - Who/Where – not shown in the title, but the **story** tells you this is for a store called Frank's Fine Furniture
    - Note the slight differences in presentation of the data –
        i. Pie Chart 1A does not provide the total revenue. If you want to calculate percentages, you will first need to calculate that total by summing each of the category revenue amounts.
        ii. Pie Chart 1B does provide the total revenue, listed below the chart title in a slightly smaller font. If you want to calculate the dollar amounts for each category, you will need to calculate these by multiplying the total revenue amount by each of the various percentages. *The total revenue could also be placed in one of two other locations:*
            1. *In a **footnote** below the pie chart*
            2. *In the **story** above the pie chart*

3. **Read the labels** and **note the units of measurement for each variable**.
    - Pie Chart 1A shows the category labels next to each of the corresponding sectors of the pie chart.
    - Pie Chart 1B does not show the category labels next to each of the corresponding sectors of the pie chart. Instead, the labels are shown in the legend on the right-hand side of the chart.

4. **Read the legend**
    - Pie Chart 1A does not need a legend, because the category labels are shown next to each of the corresponding sectors of the pie chart.
    - Pie Chart 1B uses shows the legend on the right-hand side of the chart. *This example uses color, but either greyscale or patterns (such as dots, stars, stripes or other symbols) are used frequently in lieu of color.*

5. **Get a sense of the data** without dwelling on the details. Focus on the big picture.
    - The category Sofas & Loveseats generates the most revenue for Frank's Fine Furniture.
    - The category Accent Seating generates the least revenue for Frank's Fine Furniture.
    - Several categories are similarly sized (nearly the same values / same percentages).

6. **Read any footnotes and/or explanations** – not applicable

PIE CHART 1A/1B: ANSWERS & EXPLANATIONS

Question 1: *Refer to Pie Chart 1A. The three highest-revenue categories generate approximately what percent of the total revenue at Frank's Fine Furniture?*

a) 31.3%
b) 43.8%
c) 44.3%
d) 53.6%
e) 56.2%

- Recognizing the type of question: **Part-to-total comparison**. Specifically, it's asking for the **percentage of** info about the top three categories (sum of a few parts) to the same info about all categories (a total).
- Solving the question: Use **Words to Math** and **Words First, Math Second**.
    - "The three highest-revenue categories" → examine the pie chart and carefully identify the three largest values
    - "generate approximately what percent of..." → set up a fraction. *Adjustment:* Because the phrase "the three highest-revenue categories" came <u>before</u> the phrase "what percent of", then you know that "the top three highest-revenue categories" goes in the <u>numerator</u> of your fraction.
    - "the total revenue at Frank's Fine Furniture" → this comes <u>second</u>, so it goes in the <u>denominator</u> of your fraction
    - Retrieve the necessary pieces of information from the table. Next to the fraction, place the values for each to the right. Then, simplify.

$$\frac{Three\ highest-revenue\ categories}{Total\ Revenue\ at\ Frank's\ Fine\ Furniture} = \frac{Sofas\&Loveseats + Tables + Shelving\ Units}{Sum\ of\ all\ 7\ categories}$$

$$\frac{425000+315000+272000}{425000+225000+148000+315000+272000+215000+200000} = \frac{1012000}{1800000} = 56.2\%$$

    - Choose E.
- **Behind each wrong answer choice is faulty logic**:
    - Choice A is incorrect because it is the percentage of the <u>three lowest-revenue</u> categories. If you chose this one, you may have misread the question.
    - Choice B is incorrect because it is the percentage of the <u>four lowest-revenue</u> categories. If you chose this one, you may have misread the question.
    - Choice C is incorrect because it is the percentage of the <u>first three</u> categories, starting with the highest-revenue category and reading the pie chart clockwise. If you chose this one, you may have <u>made a bad assumption</u> that the categories were arranged from largest to smallest.
    - Choice D is incorrect because it is the percentage of three categories, but not the three largest. If you chose this one, it is likely that you saw the $225,000 but did not see the $272,000. You may have read the chart too quickly or misread $2**7**2,000 as $2**1**2,000.

Question 2: *Refer to Pie Chart 1B. Approximately how much revenue did Frank's Fine Furniture generate from sales of Shelving Units and Case Goods?*

   a) $214,000
   b) $272,000
   c) $315,000
   d) $486,000
   e) $529,000

- Recognizing the type of question: **Combining values**. Specifically, it's asking for **how much** the value will be from two specific categories.
- Solving the question: Use **Words to Math** and **Words First, Math Second**.
    - "Approximately how much revenue did Frank's Fine Furniture generate..." → here, revenue is an absolute number. You will need to transform the relative numbers (percentages) into absolute numbers (amounts).
    - "from sales of Shelving Units" → 15.1%, from the pie chart
    - "and Case Goods" → 11.9%, from the pie chart
    - The value of the revenue from Shelving Units and Case Goods will be equal to the sum of the two percentages times the total revenue. Write your expression using the **Words First, Math Second** approach.
    - $Revenue\ from\ SU\ \&\ CG = (\%\ from\ SU + \%\ from\ CG) * Total\ Revenue$
    - $Revenue\ from\ SU\ \&\ CG = (15.1\% + 11.9\%) * 1800000$
    - $Revenue\ from\ SU\ \&\ CG = (27.0\%) * 1800000 = \$486,000$
    - Choose D.
    - Note: it's equally correct to calculate it a different way:
    - $(\%\ from\ SU * Total\ Revenue) + (\%\ from\ CG * Total\ Revenue)$
- **Behind each wrong answer choice is faulty logic**:
    - Choice A is incorrect because it is the percentage of revenue from <u>Case Goods</u> times the total revenue. If you chose this one, it is likely that you took the 2nd calculation approach shown above, did part of the work, and saw that your calculated result was an available answer. Make sure to structure your approach before you begin, so you do all the steps.
    - Choice B is incorrect because it is the percentage of revenue from <u>Shelving Units</u> times the total revenue. If you chose this one, it is likely that you took the 2nd calculation approach shown above, did part of the work, and saw that your calculated result was an available answer. Make sure to structure your approach before you begin, so you do all the steps.
    - Choice C is incorrect because it is the percentage of revenue from <u>Tables</u> times the total revenue. If you chose this one, you likely misread the legend.

- Choice E is incorrect because it is the percentage of revenue from <u>Tables</u> and <u>Case Goods</u> combined, times the total revenue. If you chose this one, you likely misread the legend.

Question 3: *Refer to Pie Chart 1B. Approximately how much was the median category revenue at Frank's Fine Furniture?*

   a) $214,000
   b) $225,000
   c) $257,000
   d) $315,000
   e) $425,000

- Recognizing the type of question: **Statistical interpretation**. Specifically, it's asking for the **median** value of the seven categories.
- Solving the question: Use **Words to Math** and **Words First, Math Second**.
    - "Approximately how much was the median..." → the median is the middle value, when all the values are ordered from least to greatest.
        - Logically, the order of the percentages will be the same as the order of the values, so you do not need to calculate all seven values.
        - Order the percentages from least to greatest, then find the middle percentage. That's 12.5%.
    - "category revenue" → You need to transform the relative number (percentage) into an absolute number (dollar amount).
    - Write your expression using the **Words First, Math Second** approach.
    - $Median\ Category\ Revenue = Median\ Percentage * Total\ Revenue$
    - $Median\ Category\ Revenue = 12.5\% * 1800000 = \$225,000$
    - Choose B.
- **Behind each wrong answer choice is faulty logic**:
    - Choice A is incorrect. It's just a wildcard answer to make a fifth choice available.
    - Choice C is incorrect because it is the <u>mean (average)</u> percentage in the pie chart times total revenue. If you chose this one, you may have misread the question, or mixed up the definitions of the related-but-different terms <u>median</u> and <u>mean</u>.
    - Choice D is incorrect because it is the 4th percentage in the pie chart, starting with the highest-revenue category and reading the pie chart clockwise, times total revenue. If you chose this one, you may have <u>made a bad assumption</u> that the categories were already arranged in size order.
    - Choice E is incorrect because it is the <u>maximum</u> percentage in the pie chart times total revenue. If you chose this one, you may have misread the question.

Question 4: *Refer to Pie Chart 1A.* Which of the following statements are true, based upon the information in the pie chart? Choose all that apply.

   a) The range of category revenues is $277,000.
   b) The revenue from the Tables category is more than 150% of the revenue from Décor Items.
   c) The revenue from the Sofas & Loveseats category is more than 150% more than the revenue from Décor Items.

- Recognizing the type of question: Each statement may require a different approach, so you'll need to use the same general process but different specific steps.
- Solving the question: Use **Words to Math** and **Words First, Math Second**.
- Evaluate Statement A:
    - "The range of category revenues" – **Statistical Interpretation**. The range is the maximum value minus the minimum value.
        - Maximum = $425,000
        - Minimum = $148,000
    - "is $277,000" – equals $277,000
    - Evaluate the truthfulness of the statement.
        - $Range = \$277,000$
        - $Maximum - Minimum = \$277,000$
        - $\$425,000 - \$148,000 = \$277,000$
        - $\$277,000 = \$277,000$
        - Yes, these are equal. **Statement A is true**.
- Evaluate Statement B:
    - "The revenue from the Tables category" – $315,000
    - "is more than" – set up an inequality
    - "150% of" – use the **percent of** relationship, which means **percentage times**
    - "the revenue from Décor Items" - $200,000
    - Evaluate the truthfulness of the statement.
        - $Tables\ Revenue > \frac{150}{100} * Décor\ Items\ Revenue$
        - $\$315,000 > \frac{150}{100} * \$200,000$
        - $\$315,000 > \$300,000$
        - Yes, the inequality is true. **Statement B is true**.

- Evaluate Statement C:
    - "The revenue from the Sofas & Loveseats category" - $425,000
    - "is more than" – set up an inequality
    - "150% more than" – use the **percent change** relationship, which means **(1 + percent change) times**
    - "the revenue from Décor Items" - $200,000
    - Evaluate the truthfulness of the statement.
        - $Sofas\ \&\ Loveseats\ Revenue > \left(1 + \frac{150}{100}\right) * Décor\ Revenue$
        - $\$425{,}000 > 2.5 * \$200{,}000$
        - $\$425{,}000 > \$500{,}000$
        - No, the inequality is not true. **Statement C is false**.
- **You must choose both A and B, but not C, to get the point for the question**.
- Did you catch the subtle change in the wording of Statements B and C right away? The trap for most students is the "percent more than" in the second statement.
    - You might make a bad assumption that you can use the same process to evaluate Statement C that you used to evaluate Statement B, but the wording change should prompt you to adjust your approach. Tricky!
    - You might misread Statement C and not even notice the "more" in "150% more than." Pay close attention to similar-but-different wording!

PIE CHART 2: PATIENT REFERRAL SOURCES

An orthopedic specialist receives most of her new patients via referrals from neighborhood clinics, family physicians, and community hospitals. Data concerning referrals for the current year were collected in the chart, shown below.

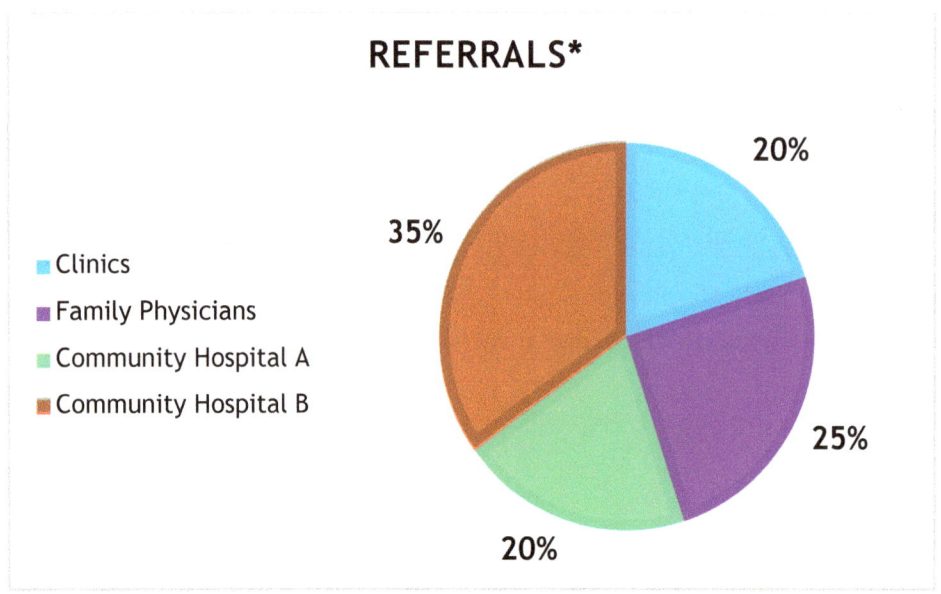

*Total patient referrals = 300

Question 1: *Refer to Pie Chart 2.* The number of patient referrals from all sources grew by approximately 12% from last year to the current year. If clinics contributed 15% of total referrals last year, then approximately how many more patients were referred by clinics in the current year than last year?

Enter your answer here

Question 2: *Refer to Pie Chart 2.* The ratio of the number of patients referred by Community Hospital A to the number of patients referred by both community hospitals is approximately:

    a) 1 to 5
    b) 4 to 11
    c) 4 to 7
    d) 11 to 4
    e) 5 to 1

PIE CHART 2: INTERPRETING THE DATA

An orthopedic specialist receives most of her new patients via referrals from neighborhood clinics, family physicians, and community hospitals. Data concerning referrals for the current year were collected in the chart, shown below.

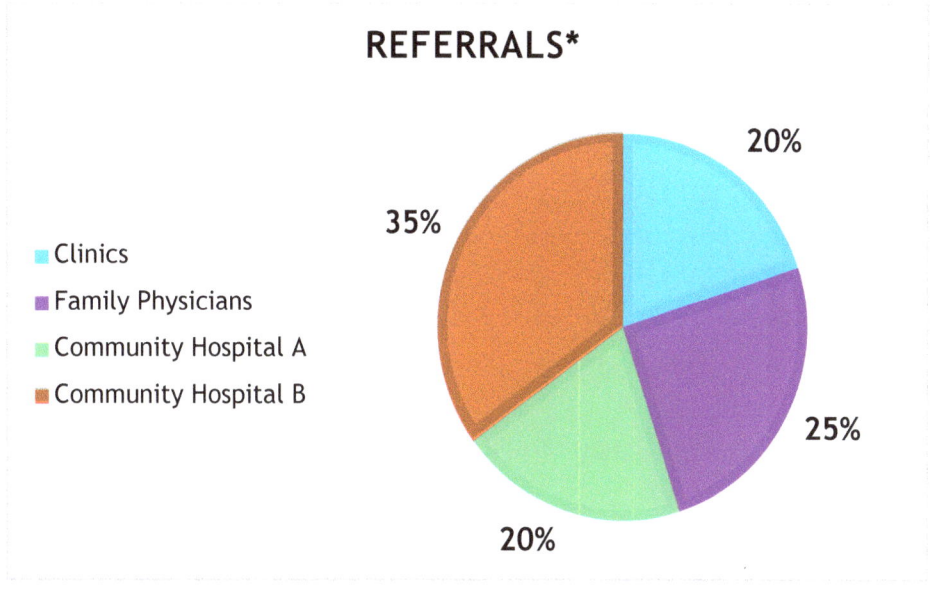

*Total patient referrals = 300

1. **Read the story** – from this, you learn that the data are about an orthopedic specialist's patient referrals.

2. **Read the title** – Referrals
    - When – the **story** simply says the data are for the current year
    - What – referrals
    - Who/Where – not shown in the **title** or the **story**

3. **Read the labels** and **note the units of measurement for each variable**.
    - There are four categories of referrals – these are the sources who send referrals to the orthopedic specialist

4. **Read the legend**
    - The legend on is on the left-hand side of the chart.
    - *This example uses color, but either greyscale or patterns (such as dots, stars, stripes, or other symbols) are used frequently in lieu of color.*

5. **Get a sense of the data** without dwelling on the details. Focus on the big picture.
    - The categories Clinics and Community Hospital B contribute an equal percentage of the total referrals.
    - Combined, the two Community Hospitals contribute more than half the total referrals.

6. **Read any footnotes and/or explanations** – the asterisk tells you there were 300 total referrals.

Off the Charts! Data Interpretation

PIE CHART 2: ANSWERS & EXPLANATIONS

Question 1: *Refer to Pie Chart 2. The number of patient referrals from all sources grew by approximately 12% from last year to the current year. If clinics contributed 15% of total referrals last year, then approximately how many more patients were referred by clinics in the current year than last year?*

Enter your answer here

- Recognizing the type of question: **Absolute comparisons about two values**. The key phrase "how many more" indicates you need an absolute number (not a relative number).
- Solving the question: Use **Words to Math** and **Words First, Math Second**.
    - "The number of patient referrals from all sources" → total referrals
    - "grew by approximately 12% from last year to the current year" → use the <u>percent change</u> formula. The percent change is 12%.
    - $Total\ Referrals\ Current = Total\ Referrals\ LY * (1 + Change)$
    - Find the information needed from the pie chart. The <u>footnote</u> indicates the total referrals in the current year = 300. The <u>question prompt</u> tells you the percent increase is 12%.
    - Plug this information into your equation and solve.
    - $300 = Total\ Referrals\ LY * (1 + 0.12)$
    - $\frac{300}{1.12} = Total\ Referrals\ LY = 267.86 = \mathbf{268}$
    - "If clinics contributed 15% of total referrals last year" → use the <u>percent of</u> formula.
    - $Clinic\ Referrals\ LY = Percent * Total\ Referrals\ LY$
    - Plug in the correct numbers into your equation and solve.
    - $Clinic\ Referrals\ LY = \frac{15}{100} * 268 = 40.2 = \mathbf{40}$
    - "then approximately how many more patients were referred by clinics in the current year than last year?" → the key phrase "how many more" means you must subtract
    - $How\ Many\ More = Clinic\ Referrals\ Current - Clinic\ Referrals\ LY$
    - Find the information needed from the pie chart. The <u>footnote</u> indicates the total referrals in the current year = 300. The <u>pie chart</u> tells you clinic referrals are 20% of the current year referrals.
    - Plug in the correct numbers into your equation and solve.
    - $How\ Many\ More = \frac{20}{100} * 300 - 40 = 60 - 40 = \mathbf{20}$
    - The final answer is 20.

Question 2: *Refer to Pie Chart 2.* The ratio of the number of patients referred by Community Hospital A to the number of patients referred by both community hospitals is approximately:

    a) 1 to 5
    b) 4 to 11
    c) 4 to 7
    d) 11 to 4
    e) 5 to 1

- Recognizing the type of question: **Part-to-part comparison**. Specifically, it's asking for the **ratio** of info about one group (a part) to the same info about a different group (a part).
- Solving the question: Use **Words to Math** and **Words First**, **Math Second**.
    - "The ratio of..." → set up the ratio as a fraction
    - "the number of patients referred by Community Hospital A" → this comes first, so it goes in the numerator of your fraction
    - "to the number of patients referred by both community hospitals" → this comes second, so it goes in the denominator of your fraction
    - $$\frac{\text{\# Referred by Community Hospital A}}{\text{\# Referred by Community Hospital A} + \text{\# Referred by Community Hospital B}}$$
    - The pie chart does not provide the numbers, so take the formula and drill down into each part.
    - $$\frac{\text{\% Referred by A} * \text{Total Referrals}}{(\text{\% Referred by A} + \text{\% Referred by B}) * \text{Total Referrals}}$$
    - Notice that Total Referrals is part of both the numerator and denominator. These cancel out.
    - Retrieve the necessary pieces of information from the pie chart.
    - $$\frac{\text{\% Referred by A}}{(\text{\% Referred by A} + \text{\% Referred by B})} = \frac{\frac{20}{100}}{\left(\frac{20}{100} + \frac{35}{100}\right)} = \frac{60}{\frac{55}{100} * 300} = \frac{60}{165} = \frac{4}{11}$$
    - Choose B, 4 to 11.
- **Behind each wrong answer choice is faulty logic**:
    - Choice A is incorrect. It is the ratio of the number of patients referred by Community Hospital A to the number of referrals from all sources. If you chose this answer, you may have read too quickly or made an assumption.
    - Choice C is incorrect. It is the ratio of the number of patients referred by Community Hospital A to the number of patients referred by Community Hospital B. If you chose this answer, you may have reading too quickly or made an assumption.
    - Choice D is incorrect because it is the reciprocal of the correct answer. Be cautious with ratios so that you put the right part in the right place!
    - Choice E is incorrect for two reasons. It is the reciprocal of answer choice A. See that explanation above.

CHAPTER 5 LINE CHARTS

CONCEPT REVIEW

A **line chart** is one of the more common types of data visualizations. **Line charts** are most often used to show **trends** over time.

A line chart can have:

- A single X-axis. When the goal is to show **trends** or otherwise present **time series data**, time will be the **independent variable** shown on the X-axis.
- A single or double Y-axis.
- One or more **dependent variables** graphed as separate lines on the same line graph.

Recall that the **slope** of a line represents the **rate** of change. If you are asked to find the average change across multiple years, you do not need to calculate each change and then average these changes. Instead, you can simply calculate the average change using:

$$Average\ Change\ per\ Year = \frac{Final\ Year\ Value - First\ Year\ Value}{Number\ of\ Years\ Elapsed}$$

Both methods will get you the same result, but the first method is too slow; the second method is faster.

On a **dual-line** or **multiple-line** graph, the different lines may be distinguished using color, symbols, and/or textures (such as dots, dashes, or doubled lines). Refer to the **legend** or **key** as needed.

Sometimes, a graph may feature a **broken Y-axis**, with a hashed line to indicate that the graph does not start at the **origin**. Instead, the Y-axis may start at 500 or 200,000 or some other non-zero number. Watch out for this type of graph, because the trends may appear more or less severe than they actually are when the graph starts at a non-zero value.

You should also watch out for graphs with numbers that look like they could be percentages but are not, or ones that look like numbers but are labeled with the word "percent."

Let's look at a few examples.

PRACTICE SETS

LINE CHART 1: STUDENT SICK DAYS

During the 2010-2017 school years, a teacher tracked the number of total sick days her students took for various illnesses.

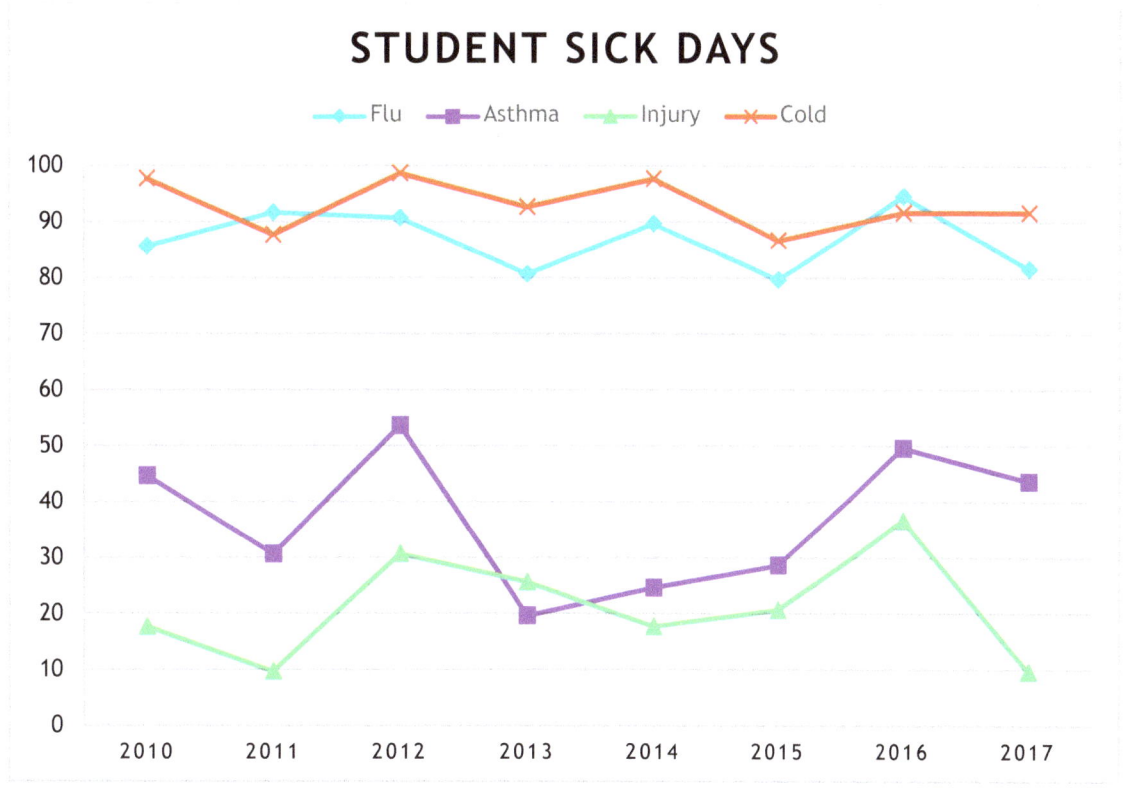

Question 1: *Refer to Line Chart 1.* In what year did the greatest number of sick days due to injury occur?

   a) 35
   b) 37
   c) 2012
   d) 2016
   e) 2017

Question 2: *Refer to Line Chart 1.* Between which years did the greatest increase in the number of sick days due to asthma occur?

   a) 2011 to 2012
   b) 2012 to 2013
   c) 2013 to 2014
   d) 2014 to 2015
   e) 2015 to 2016

Off the Charts! Data Interpretation

LINE CHART 1: INTERPRETING THE DATA

During the 2010-2017 school years, a teacher tracked the number of total sick days her students took for various illnesses.

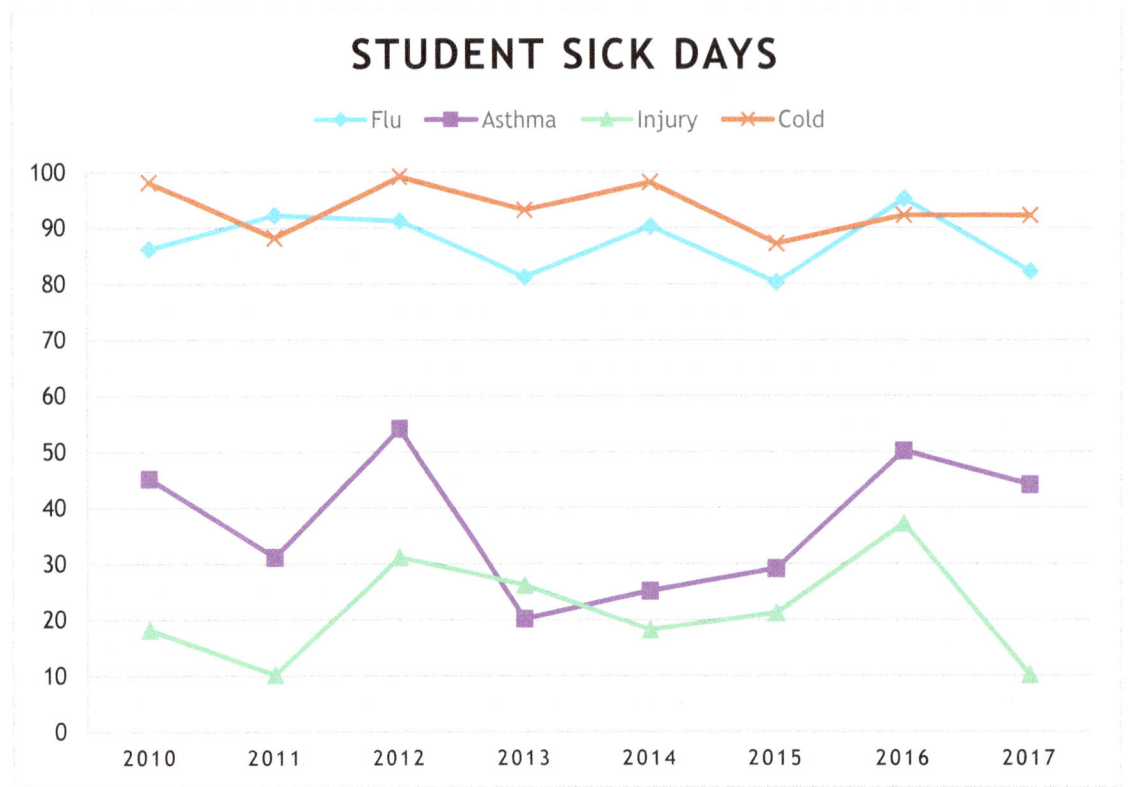

1. **Read the story** – from this, you learn that the data are about one teacher's classes, calculating the number of sick days for four different types of illnesses, over an eight-year period from 2010 through 2017.

2. **Read the title** – Student sick days

    - When – not shown in the title, but the **story** tells you this is for the period from 2010 to 2017. The **X-axis labels** confirm that this is for the period from 2010 to 2017.
    - What – student sick days
    - Who/Where – not shown in the title, but the **story** tells you this is for an unnamed teacher's class.

3. **Read the labels** and **note the units of measurement for each variable**.

    - The **X-axis** is not labeled, but you can logically conclude from the **story** that these are the **years**, because the numbers match up.
    - The **Y-axis** is not labeled, but you can logically conclude that the **number of student sick days** are plotted along the Y-axis because sick days would depend on the year, and you should plot **dependent variables** on the Y-axis.

4. **Read the legend**

- The **legend** is placed below the title but above the line chart.
- The **legend** uses both color and symbols to make it easier to keep straight what's what.

5. **Get a sense of the data** without dwelling on the details. Focus on the big picture.
    - The two illnesses leading to the greatest number of sick days are **Flu** and **Cold**.
    - The two illnesses leading to the fewest number of sick days are **Asthma** and **Injury**.
    - There is greater variation in the number of sick days for **Asthma** and **Injury** than for the other illnesses, because the lines are more jagged.

6. **Read any footnotes and/or explanations** – not applicable

LINE CHART 1: ANSWERS & EXPLANATIONS

Question 1: *Refer to Line Chart 1.* In what year did the greatest number of sick days due to injury occur?

a) 35
b) 37
c) 2012
d) 2016
e) 2017

- Recognizing the type of question: **Statistical interpretation**. Specifically, it's asking for the **year when** the **greatest number** of sick days occurred. Look for the maximum value (highest point) on the line related to injury.

- Solving the question: This one is straightforward.
    - Check the legend – the line with the number of sick days due to injury is the green line with triangles.
    - Look for the highest point on the line – the one that's approximately 35 or 40.
    - Go straight down from this point to the X-axis – that's **2016**.
    - The year with the greatest number of sick days **due to injury** occurred in **2016**.
    - Choose D.

- **Behind each wrong answer choice is faulty logic**:
    - Choice A is incorrect. It's a good estimate for the <u>number of sick days</u> due to injury, but not for the <u>year in which this event occurred</u>. Right answer to the wrong question.
    - Choice B is incorrect. It's a good estimate for the <u>number of sick days</u> due to injury, but not for the <u>year in which this event occurred</u>. Right answer to the wrong question.
    - Choice C is incorrect, because it's the year in which the greatest number of sick days due to <u>asthma</u> occurred. Right answer to the wrong question.
    - Choice E is incorrect. It's the year in which the <u>greatest change in the number of sick days</u> due to injury occurred. Right answer to the wrong question.

Question 2: *Refer to Line Chart 1.* Between which years did the greatest increase in the number of sick days due to asthma occur?

   a) 2011 to 2012
   b) 2012 to 2013
   c) 2013 to 2014
   d) 2014 to 2015
   e) 2015 to 2016

- Recognizing the type of question: **Statistical interpretation**. Specifically, it's asking for the **year when** the **greatest increase** in the number of sick days due to asthma occurred.
    - The **greatest change** happens when the slope is the steepest
    - The **greatest increase** is the steepest upward sloping portion of the line
    - You may be able to determine this visually; try that before calculating exact answers.
- Solving the question: This one is straightforward.
    - Check the legend – the line with the number of sick days due to asthma is the purple line with squares.
    - Look for the steepest upward sloping portion of the line – 2011 to 2012 and 2015 to 2016 seem like good candidates.
    - On a graph this small, it's hard to discern the exact values for those years. That's a good hint you're not supposed to need the exact values.
    - Instead, use **reasoning**.
        - The change from 2011 to 2012 appears to be about 5 of the gridlines (2 dark gridlines and 3 light gridlines).
        - The change from 2015 to 2016 appears to be about 4 of the gridlines (2 dark gridlines and 2 light gridlines).
        - Thus, you can conclude that the greatest change occurred from **2011 to 2012**.
    - Choose A.
- **Behind each wrong answer choice is faulty logic**:
    - Choice B is incorrect. In this time period, the <u>greatest change</u> in the number of sick days due to asthma occurred, but it was a <u>decrease</u>. This is the right answer to the wrong question. If you chose this one, you may have misread the question.
    - Choice C is incorrect. It's just a filler answer.
    - Choice D is incorrect. It's just a filler answer.
    - Choice E is almost right, but still incorrect. See the notes above about counting the gridlines when it is hard to determine exact values from a line graph.

Off the Charts! Data Interpretation

LINE CHART 2: COMPANY FINANCIAL PERFORMANCE

An investor is contemplating purchase of a company with the following financial performance data for the years 2010 through 2015.

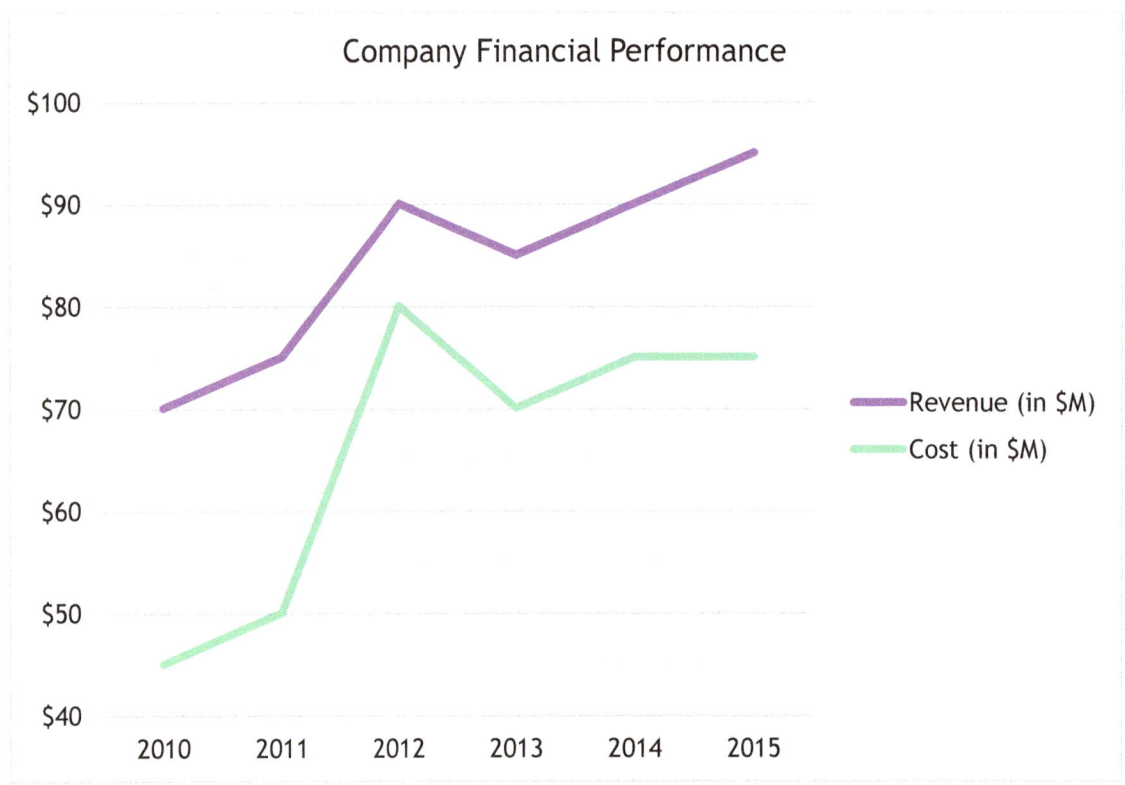

Question 1: *Refer to Line Chart 2.* By what amount did the company's revenue change from 2011 to 2012?

    a) $15 million decrease
    b) $15 million increase
    c) $30 million increase

Question 2: *Refer to Line Chart 2.* Which of the following statements are true? Select all that apply.

    a) The company's profits declined from 2012 to 2013.
    b) The company's revenue declined in only one year.
    c) The company's costs stayed the same from 2014 to 2015.
    d) The company's profits were lowest in 2012.

Question 3: *Refer to Line Chart 2.* Based upon the information in the graph, compare the following quantities:

Quantity A	Quantity B
The percent increase in costs from 2013 to 2014	The percent increase in costs from 2010 to 2011

a) Quantity A is greater
b) Quantity B is greater
c) Quantity A and Quantity B are equal
d) Cannot be determined

LINE CHART 2: INTERPRETING THE DATA

An investor is contemplating purchase of a company with the following financial performance data for the years 2010 through 2015.

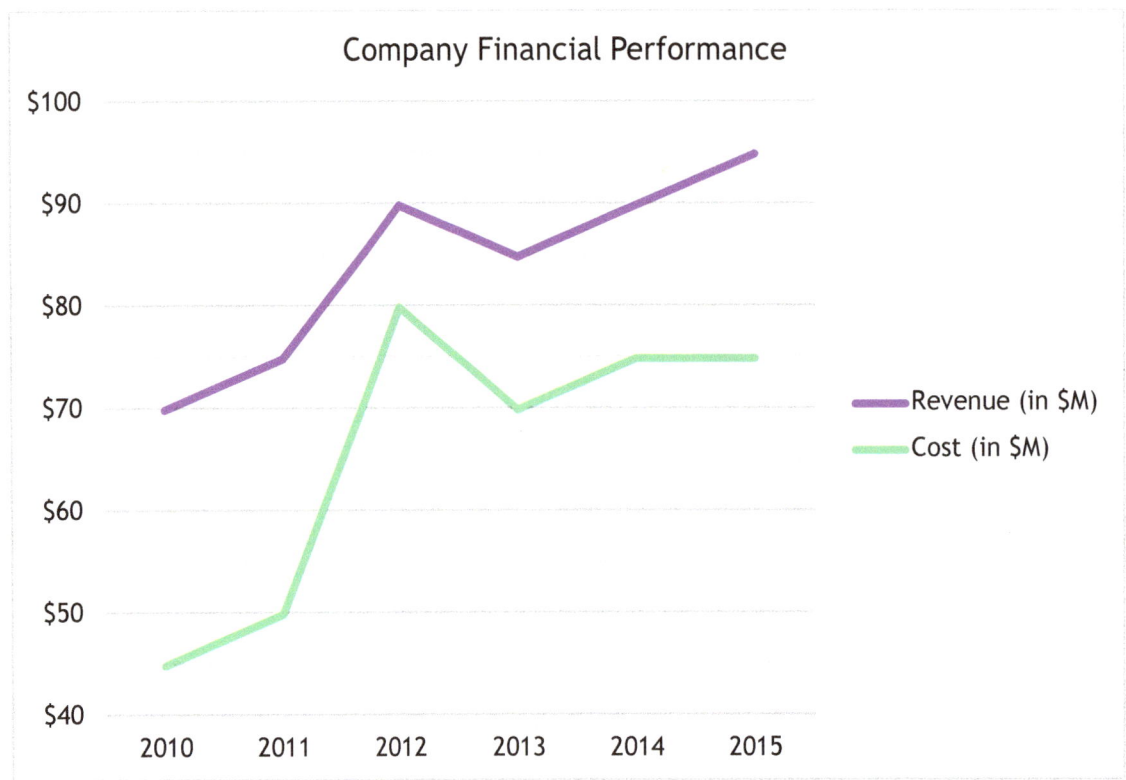

1. **Read the story** – from this, you learn that the data are about a company's financial performance for 2010 through 2015.

2. **Read the title** – Company financial performance
   - When – not shown in the title, but the **story** tells you this is for the period from 2010 to 2015. The **X-axis labels** confirm that this is for the period from 2010 to 2015.
   - What – financial performance
   - Who/Where – both the **title** and the **story** tell you this is for an unnamed company.

3. **Read the labels** and **note the units of measurement for each variable**.
   - The **X-axis** is not labeled, but you can logically conclude from the **story** that these are the **years**, because the numbers match up.
   - The **Y-axis** is not labeled, but you can logically conclude that the **dollar amounts** are plotted along the Y-axis based on the units of the **Y-axis labels**.
   - The **Y-axis** does not start at zero (the origin).

4. **Read the legend**
    - The **legend** is placed to the right of the line chart.
    - The **legend** uses color to make it easier to keep straight what's what. From the legend, you learn that the specific type of financial information included in the graph pertains to Revenue and Cost. The **parenthetical notation** (in $M) means the information on the Y-axis should be interpreted in millions of dollars.
    - For example, in 2010, the purple line crosses the gridline marked $70. Combining information from the **legend**, the **parenthetical notation**, and the line chart tells you the company's revenue in 2010 was $70 million.

5. **Get a sense of the data** without dwelling on the details. Focus on the big picture.
    - For each year depicted, revenue was greater than cost. The purple line is always above the green line.
    - Both revenue and cost generally increased over time.
    - Cost increased significantly from 2011 to 2012.
    - The cost line reaches a peak in 2012. In other words, the **maximum** cost occurred in 2012.
    - Recall that revenue and cost are the two elements of profit. Thus, you should anticipate questions which require you to calculate either the amount of profit or the percentage profit.

$$Profit\ Amount\ =\ Revenue\ -\ Cost$$

$$Profit\ Percentage\ =\ \frac{(Revenue\ -\ Cost)}{Cost}$$

6. **Read any footnotes and/or explanations** – not applicable

Off the Charts! Data Interpretation

LINE CHART 2: ANSWERS & EXPLANATIONS

Question 1: *Refer to Line Chart 2. By what amount did the company's revenue change from 2011 to 2012?*

a) $15 million decrease
b) $15 million increase
c) $30 million increase

- Recognizing the type of question: **Absolute comparison**. Specifically, it's asking for the **amount** of the change in revenue from 2011 to 2012.
- Solving the question: Use **Words to Math** and **Words First, Math Second**.
    - "By what amount" → solve for a number, not a ratio or percentage
    - "did the company's revenue" → check the legend – the purple line represents revenue
    - "change" → find the difference using subtraction
    - "from 2011" → this is the original amount
    - "to 2012" → this is the new amount
    - *Amount of Change* = *New Amount 2012* – *Original Amount 2011*
    - Retrieve the necessary pieces of information from the line graph. Enter these below your formula, in their corresponding places.
    - *Amount of Change* = *New Amount 2012* – *Original Amount 2011*
    - *Amount of Change* = $90 million – $75 million
    - *Amount of Change* = $15 million
    - Because the sign is positive, interpret this as a $15 million increase.
    - Choose B.
- **Behind each wrong answer choice is faulty logic**:
    - Choice A is incorrect. You could arrive at this answer in either of two ways. First, if you selected the right information about revenue from the graph, but you reversed the order of the two numbers in the equation, then you would have incorrectly assumed the sign was negative. Second, if you assumed the question was about profit, and you calculated the change in profits. The change in profits is a $15 million decrease, so choice A is the <u>right answer to the wrong question</u>.
    - Choice C is incorrect. It is the change in <u>cost</u> during this time period. Thus, this is the right answer to the wrong question. If you chose this answer, you may have misread the question or misread the line graph and its legend.

Question 2: *Refer to Line Chart 2.* Which of the following statements are true? Select all that apply.

   a) The company's profits declined from 2012 to 2013.
   b) The company's revenue declined in only one year.
   c) The company's costs stayed the same from 2014 to 2015.
   d) The company's profits were lowest in 2012.

- Recognizing the type of question: Each statement may require a different approach, so you'll need to use the same general process but different specific steps.
- Solving the question: Use **Words to Math** and **Words First**, **Math Second**.
- Evaluate Statement A:
    - "The company's profits" → Check the legend. There is no line for profit. Recall that profit is the <u>difference</u> between revenue and cost.
    - "declined" → decreased
    - "from 2012 to 2013" → compare these two years
    - Evaluate the truthfulness of the statement.
        - Determine profit in 2012
        - $Profit\ in\ 2012 = Revenue\ in\ 2012 - Cost\ in\ 2012$
        - $Profit\ in\ 2012 = \$90\ million - \$80\ million$
        - $Profit\ in\ 2012 = \$10\ million$
        - Determine profit in 2013
        - $Profit\ in\ 2013 = Revenue\ in\ 2013 - Cost\ in\ 2013$
        - $Profit\ in\ 2013 = \$85\ million - \$70\ million$
        - $Profit\ in\ 2013 = \$15\ million$
        - Profit did not decline. Profit increased.
        - Shortcut: Profit, on this line graph, is represented by the vertical distance between the revenue and cost points for any given year. Visual inspection of the graph shows that the vertical distance between revenue and cost is <u>clearly larger in 2013 than in 2012</u>. This visual information is enough to conclude that statement A is false – start to recognize when <u>estimating a good-enough answer</u> works instead of <u>calculating a precise answer</u>, because estimating can save you significant time on the test.
        - **Statement A is false**.
- Evaluate Statement B:
    - "The company's revenue declined" → Check the legend. Go to the purple line for revenue.
    - "in only 1 year" → one time

Off the Charts! Data Interpretation

- o Shortcut: Count the number of times that the purple line, which represents revenue, slopes downward. This happens once.
- o **Statement B is true**.
- Evaluate Statement C:
    - o "The company's costs stayed the same" → Check the legend. Go to the green line for costs.
    - o "from 2014 to 2015" → Compare these two years.
    - o Shortcut: Determine whether the green line, which represents cost, is flat or not. Because the slope of the green line from 2014 to 2015 is flat (zero slope), you know that costs stayed the same.
    - o **Statement C is true**.
- Evaluate Statement D:
    - o "The company's profits" → Check the legend. There is no line for profit. Recall that profit is the difference between revenue and cost.
    - o "were lowest in 2012" → In 2012, profit was lower than in any other year.
    - o Shortcut: Profit, on this line graph, is represented by the vertical distance between the revenue and cost points for any given year. Visually inspect the graph to determine when the revenue and cost points for any single year are closest together.
    - o The revenue and cost lines are closest together in 2012, so yes, this is the year when profits were lowest.
    - o **Statement D is true**.
- **You must choose B, C, and D, but not A, to get the point for the question**.

Question 3: *Refer to Line Chart 2.* Based upon the information in the graph, compare the following quantities:

Quantity A

The percent increase in costs from 2013 to 2014

Quantity B

The percent increase in costs from 2010 to 2011

a) Quantity A is greater
b) Quantity B is greater
c) Quantity A and Quantity B are equal
d) Cannot be determined

- Recognizing the type of question: For questions which ask you to compare two quantities (typically found on the GRE®), you may need to translate one or two statements from words to math. Sometimes, the second quantity is simply a number (such as zero or 20) which requires no translation.
- Solving the question: Use **Words to Math** and **Words First, Math Second**.
- Evaluate Quantity A:
    - "The percent increase" = use the percent change formula
    - $Percent\ Change = \frac{(New - Original)}{Original}$
    - "in costs" = obtain cost information from the line chart. This is the green line.
    - "from 2013" = cost in 2013 is the original
    - "to 2014" = cost in 2014 is the new
    - $Percent\ Change = \frac{(Cost\ in\ 2014\ -\ Cost\ in\ 2013)}{Cost\ in\ 2013}$
    - $Percent\ Change = \frac{(75\ million\ -\ 70\ million)}{70\ million} = \frac{5\ million}{70\ million} = 0.07\ or\ 7\%$
- Evaluate Quantity B:
    - "The percent increase" = use the percent change formula
    - $Percent\ Change = \frac{(New - Original)}{Original}$
    - "in costs" = obtain cost information from the line chart. This is the green line.
    - "from 2010" = cost in 2010 is the original
    - "to 2011" = cost in 2011 is the new
    - $Percent\ Change = \frac{(Cost\ in\ 2011\ -\ Cost\ in\ 2010)}{Cost\ in\ 2010}$

- $Percent\ Change = \frac{(50\ million - 45\ million)}{45\ million} = \frac{5\ million}{45\ million} = 0.11\ or\ 11\%$
  - **Quantity B is greater**.
  - **Choose B**.

The most common incorrect answer for this GRE-style quantity comparison question would be to choose C, that the two quantities are equal. If you chose C, you may have rushed through your work and simply compared the <u>amount</u> of the change, rather than the <u>percentage</u> change.

CHAPTER 6 BAR & COLUMN CHARTS

CONCEPT REVIEW

Bar charts and **column charts** are among the more common types of data visualizations. **Bar charts** and **column charts** are most often used to show **absolute comparisons** between categories.

A bar chart or column chart can have:

- A single X-axis.
- A single Y-axis.
- One or more than one **dependent variable**, graphed as separate lines on the same line graph.

The key difference between the two is that bar charts are oriented horizontally, and column charts are oriented vertically.

A **stacked column** chart is a special variation with **stacked values**, for each column.

- Provides extra information – because it visually conveys both the **total value** for each X-variable and the **breakdown by sub-category** within the X-variable.
- Helps you to visually see the relative mix among **sub-categories** and how each contributes toward the total value.
- Needs a **legend** or **data labels**. To find the sub-category quantities, look at the legend, find the right color or markings for the sub-category of interest, then find the values at the top and bottom of that segment of the column. Take the difference of these two values to determine the value for that sub-category.

Let's look at a few examples.

Off the Charts! Data Interpretation

PRACTICE SETS

COLUMN CHART 1: MUSICAL INSTRUMENTS PLAYED

The City Parks & Recreation department is planning to introduce music classes for members of a certain community. To manage its budget wisely, the department conducted a survey of 200 people to identify for which musical instruments there would be the most demand.

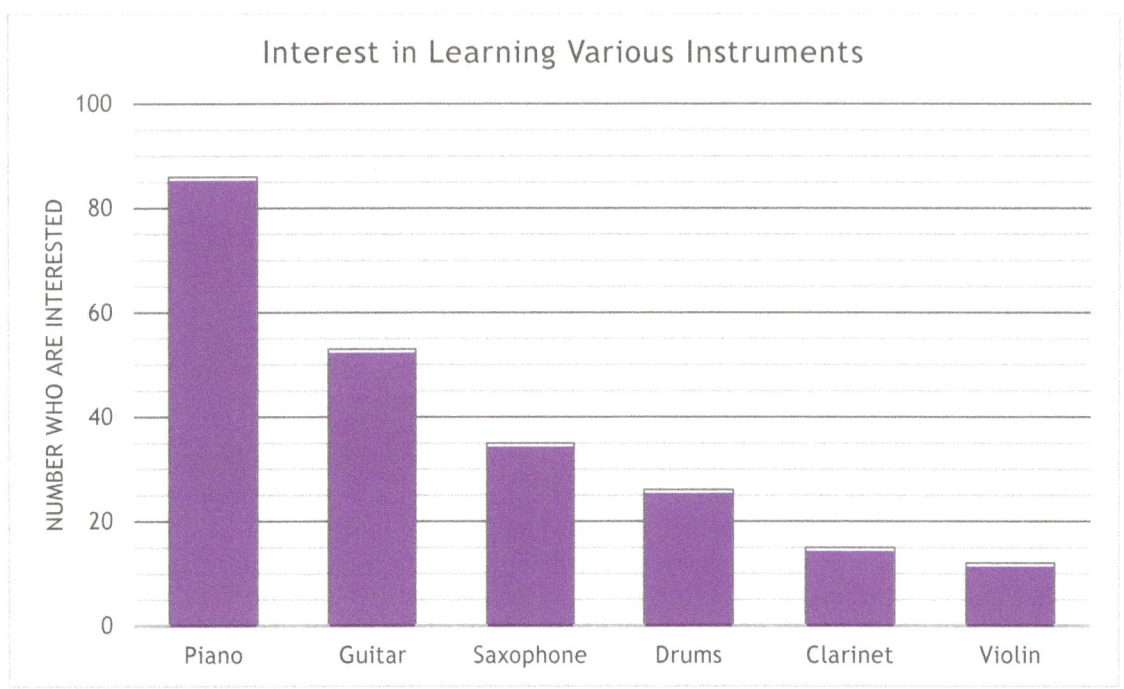

Question 1: *Refer to Column Chart 1.* Approximately what percent of survey respondents are interested in learning the saxophone?

   a) 13%
   b) 18%
   c) 26%
   d) 35%
   e) 52%

Question 2: *Refer to Column Chart 1.* Which of the following statements must be true, based upon the information in Column Chart 1? Choose all that apply.

   a) The number of people who are interested in learning guitar is nearly double the number of people interested in learning drums.
   b) All survey respondents are interested in playing at least one instrument.
   c) Some survey respondents must have reported interest in playing more than one instrument.

COLUMN CHART 1: INTERPRETING THE DATA

The City Parks & Recreation department is planning to introduce music classes for members of a certain community. To manage its budget wisely, the department conducted a survey of 200 people to identify for which musical instruments there would be the most demand.

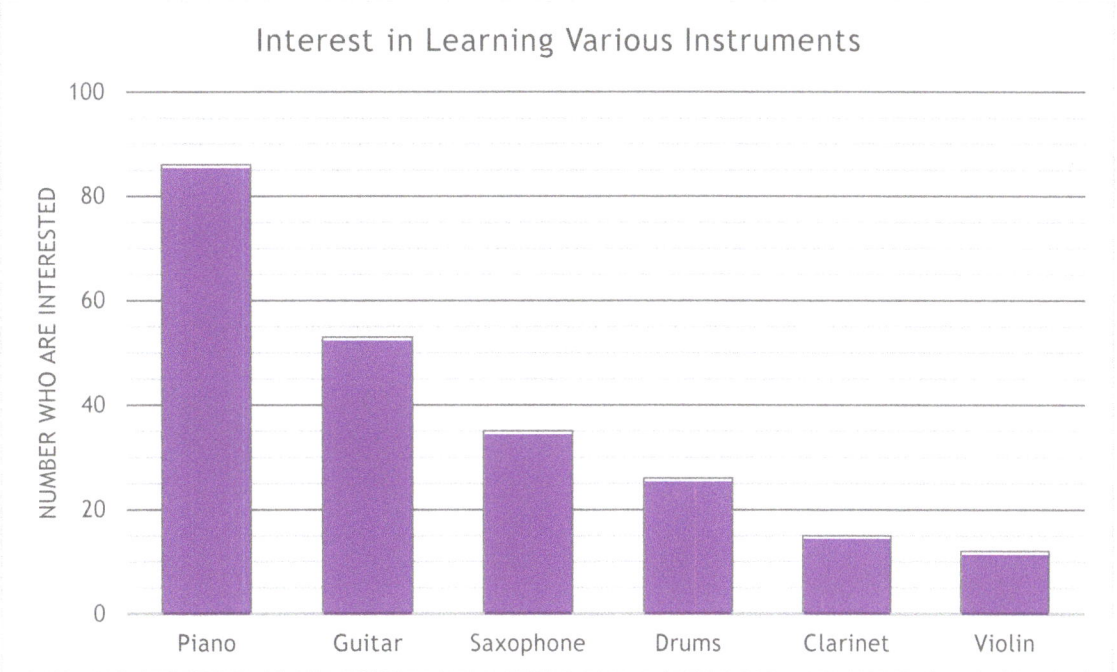

1. **Read the story** – from this, you learn that the data are from a survey about the demand for music instruction, by type of instrument. The survey population is 200 people.

2. **Read the title** – Interest in learning various instruments

    - When – not shown in the title or the story
    - What – interest in learning various instruments
    - Who/Where – not shown in the title, but the **story** tells you this is for a certain community.

3. **Read the labels** and **note the units of measurement for each variable**.

    - The **X-axis** as a whole is not labeled, but the independent variable is clearly the type of instrument because each column is labeled individually. An appropriate X-axis label could be "type of instrument" or "top six instruments" (assuming that were true).
    - The **Y-axis** is labeled, showing the number who are interested. This axis is scaled from 0 to 100, which means it could be <u>easy to make a mistake</u> and assume these are <u>percentages</u>, but the label confirms these are <u>numbers</u> of people. The major gridlines are numbered in increments of 20. The minor gridlines are not numbered but you can quickly figure out that these are in increments of 5.

4. **Read the legend** – none needed

Off the Charts! Data Interpretation

5. **Get a sense of the data** without dwelling on the details. Focus on the big picture.
    - The instruments are presented in the column chart with the highest-interest instrument listed first, and the rest shown in **descending order** of level of interest.
    - The number of people who are interested in clarinet and who are interested in violin are the closest in value.
6. **Read any footnotes and/or explanations** – not applicable

COLUMN CHART 1: ANSWERS & EXPLANATIONS

Question 1: *Refer to Column Chart 1. Approximately what percent of survey respondents are interested in learning the saxophone?*

a) 13%
b) 18%
c) 26%
d) 35%
e) 52%

- Recognizing the type of question: **Part-to-total comparison**. Specifically, it's asking for the **percent of** the **survey respondents** who want to learn the saxophone.
- Solving the question: Use **Words to Math** and **Words First**, **Math Second**.
    - "What percent of" → this means you need a **variable** times something
    - "survey respondents..." → this follows "of..." which means you multiply our variable percentage by the number of survey respondents
    - "are" → all forms of the verb "to be" mean "equals"
    - "interested in learning the saxophone" → find this value from the chart.
    - Retrieve the necessary pieces of information from the table. Next to the fraction, place the values for each to the right. Then, calculate the percentage.
    - $What\ Percent * Survey\ Respondents = Interested\ in\ Saxophone$
    - $P * 200\ Respondents = 35\ Saxophone$
    - $P = \frac{35\ Saxophone}{200\ Respondents} = \mathbf{17.5\%}$
    - Choose B.
- **Behind each wrong answer choice is faulty logic**:
    - Choice A is incorrect. Both ways of arriving at this choice assume you pulled the wrong data point from the chart.
    - Choice C is incorrect because it uses the value for guitars, not saxophones. It does however correctly combine the numeric information from the chart with the information in the story that 200 people responded to the survey.
    - Choice D is incorrect because it is the result of assuming the Y-axis represents percentages instead of numbers, and/or not combining the numeric information from the chart with the information in the story that 200 people responded to the survey.
    - Choice E is incorrect for two reasons. First, because it uses the value for guitars, not saxophones. Second, it is the result of assuming the Y-axis represents percentages instead of numbers, and/or not combining the numeric information with the information in the story that 200 people responded to the survey.

Off the Charts! Data Interpretation

Question 2: *Refer to Column Chart 1.* Which of the following statements must be true, based upon the information in Column Chart 1? Choose all that apply.

   a) The number of people who are interested in learning guitar is nearly double the number of people interested in learning drums.
   b) All survey respondents are interested in playing at least one instrument.
   c) Some survey respondents must have reported interest in playing more than one instrument.

- Recognizing the type of question: Each statement may require a different approach, so you'll need to use the same general process but different specific steps.
- Solving the question: Use **Words to Math** and **Words First, Math Second**.
- Evaluate Statement A:
    - "Number of people interested in learning guitar" → about 52 or 53
    - "is nearly double" → approximately equals 2 times (something)
    - "The number of people interested in learning drums" → about 25 or 26
    - $\# Guitar \cong 2 * \# Drums$
    - Evaluate the truthfulness of the statement.
        - $52 \text{ or } 53 \cong 2 * (25 \text{ or } 26)$
        - Yes, these are approximately equal. **Statement A is true**.
- Evaluate Statement B:
    - "All survey respondents" → 200 people. Pay attention to "risky words" such as *all*, *every*, *none*, and *never*.
    - "Are interested in playing at least one instrument" → This phrase would imply that no respondents said they were not interested in any of the instruments. Can you logically conclude that? Why or why not?
        - Does anything in the **story** allow you to conclude that every respondent chose at least one instrument? No.
        - Does anything in the **chart data** allow you to conclude that every respondent chose at least one instrument?
            - The total number of responses is greater than the total number of people, so it is possible to assume <u>some</u> people are interested in more than one instrument.
            - Because you have no information about how many people reported interest in more than one instrument, you cannot conclude that there were no "none of the above" type responses.
    - Evaluate the truthfulness of the statement.
        - Therefore, Statement B could be true, but is not necessarily true. **Do not select Statement B**.

- Evaluate Statement C:
    - "Some survey respondents" → A non-zero number
    - "must have" → Make sure you know the logical difference between must, could, cannot, and any other helping verbs that appear in this type of statement.
    - "reported interest in playing more than one instrument" → The total number of responses (approx. 220-230) is greater than the total number of people (200), so you can conclude that <u>some</u> people must be interested in more than one instrument.
        - Evaluate the truthfulness of the statement.
        - **Statement C is true.**
- **You must choose both A and C, but not B, to get the point for the question.**

Off the Charts! Data Interpretation

COLUMN CHART 2: NEW VEHICLE SALES AT A DEALER

A car dealer has tracked new vehicle sales over the past few years and wishes to understand the chart provided by the dealership sales manager.

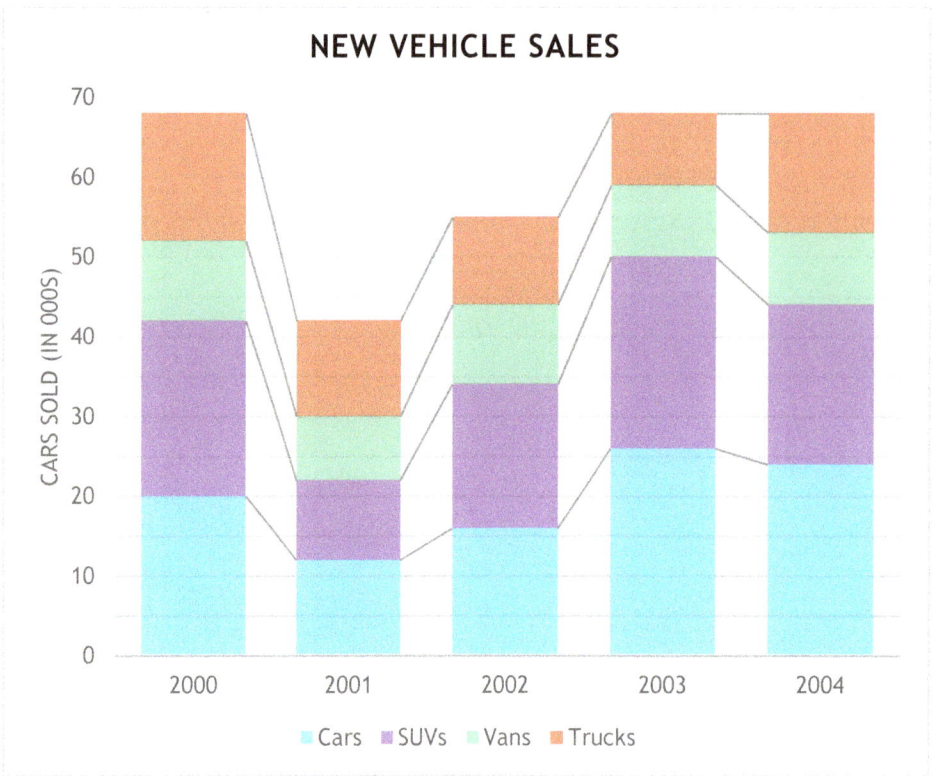

Question 1: *Refer to Column Chart 2.* In what year were the fewest trucks sold?

    a) 2000
    b) 2001
    c) 2002
    d) 2003
    e) 2004

Question 2: *Refer to Column Chart 2.* In 2001, total vehicle sales had changed by approximately what percent from the prior year?

    a) 24%
    b) 38%
    c) 47%
    d) 62%
    e) 75%

COLUMN CHART 2: INTERPRETING THE DATA

A car dealer has tracked new vehicle sales over the past few years and wishes to understand the chart provided by the dealership sales manager.

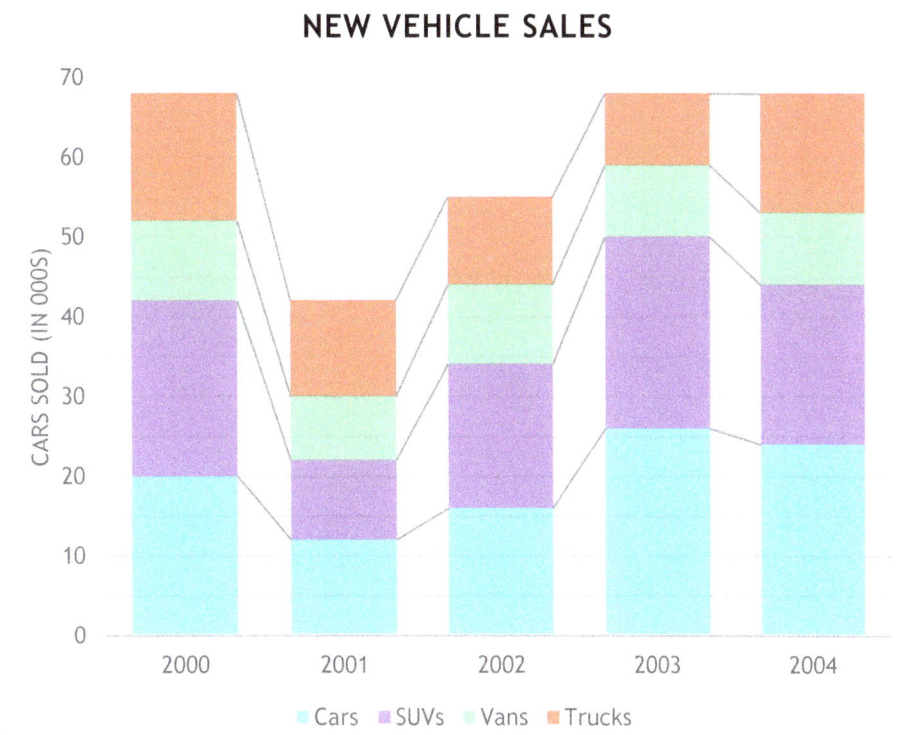

1. **Read the story** – From this, you learn that the dealership sales manager has tracked vehicle sales over several years.

2. **Read the title** – New vehicle sales

    - When – not shown in the title or the story. Instead, it is indicated by the years listed along the X-axis.
    - What – new vehicle sales.
    - Who/Where – not shown in the title, but the **story** tells you this is for a certain car dealership.

3. **Read the labels** and **note the units of measurement for each variable**.

    - The **X-axis** as a whole is not labeled, but the independent variable is clearly the year as each column is labeled individually. In this example, the year could mean the "model year" of the car or the "calendar year" in which the new vehicles were sold. By combining the information in the story and on the X-axis with a little common sense, you can conclude that an appropriate X-axis label could be "sales year."
    - The **Y-axis** is labeled, showing the number of cars sold. The note in parentheses (in 000s) means that a value of 20 really means 20,000 vehicles were sold in that year. This axis is scaled from 0 to 70, which means the vehicle sales varied between 0 and 70,000 vehicles in any given year.

Off the Charts! Data Interpretation

4. **Read the legend** – blue = cars; purple = SUVs; green = vans; orange = trucks.
    - From this information and the previous steps, you could modify the chart title to: "New vehicle sales, by type, for the sales years 2000-2014"
    - You can also see that both the total number of vehicles changed from year to year. The number of each type of vehicle also changed from year to year.

5. **Get a sense of the data** without dwelling on the details. Focus on the big picture.
    - During the 5-year period shown in the stacked column chart, total vehicle sales were lowest in the year 2001.
    - The total vehicle sales were approximately the same in 2000, 2003 and 2004. *Look at the overall height of the bar.*
    - The mix of types sold differed during those years. *Look at the height of the different colored sections.*
    - SUV sales varied quite a bit from year to year. You can see the diagonal/slanted connector lines at the top and bottom of each purple section getting either closer together (as in 2000 to 2001) or further apart (as in 2001 to 2002).
    - Note: Connector lines may or may not be present on this type of graph. You can draw your own, though.
        i. When the connector lines converge (come together), the value for that category has decreased.
        ii. When the connector lines diverge (grow apart), the value for that category has increased.

6. **Read any footnotes and/or explanations** – not applicable

COLUMN CHART 2: ANSWERS & EXPLANATIONS

Question 1: *Refer to Column Chart 2.* In what year were the fewest trucks sold?
   a) 2000
   b) 2001
   c) 2002
   d) 2003
   e) 2004

- Recognizing the type of question: **Statistical interpretation**. Specifically, it's asking for the **year when** the **smallest number** of <u>trucks</u> occurred.
- Solving the question:
    - Go to the legend first. What color is associated with trucks? Orange.
    - Scan the stacked column chart to see if you can <u>visually determine</u> which year has the smallest number of trucks sold.
    - Shortcut: When gridlines are present, you may find it **helpful to count from gridline to gridline crossed** (or box to box if you have both horizontal and vertical gridlines) by the orange section of each column. *This is less effortful than trying to read each upper and lower numeric value and perform subtraction for the five different years.*
        - 2000: 2 whole and 2 partial
        - 2001: 2 whole and 1 partial
        - 2002: 2 whole and 1 partial
        - 2003: 1 whole and 2 partial
        - 2004: 2 whole and 2 partial
        - From this, you can already eliminate 2000 and 2004 from the list. Visual inspection shows these two years have the <u>most</u> trucks sold, and you are looking for the year in which the <u>least</u> trucks sold.
        - Refine your estimation by half-boxes or quarter-boxes for the years 2001, 2002, and 2003.
        - Focus on 2003 – do those 2 partial boxes appear to be enough to combine into one full box? No. Then 2003 must be the year with the fewest trucks sold.
    - If you are not comfortable estimating, you can also calculate each value.
    - Choose D.
- **Behind each wrong answer choice is faulty logic**:
    - *Choice B is incorrect. It's the year in which the fewest <u>total vehicles</u> were sold. If you chose this one, it is likely you misread the question.*
    - *The other answers provided simply match up to each of the years shown in the column chart.*

Off the Charts! Data Interpretation

Question 2: *Refer to Column Chart 2. In 2001, total vehicle sales had changed by approximately what percent from the prior year?*

   a) 24%
   b) 38%
   c) 47%
   d) 62%
   e) 75%

- Recognizing the type of question: **Relative comparison**. Specifically, it's asking for the **percent change** in the **total number** of vehicles.
    - The question mentions 2001, but you must carefully interpret the wording – should you compare 2001 to either 2000 or to 2002?
    - The phrasing "from the prior year" tells you that you should compare 2001 to 2000, because "prior" and "before" are synonymous.
- Solving the question: Use **Words to Math** and **Words First**, **Math Second**.
    - "in 2001, total vehicle sales" → what you are focused on
    - "had changed by approximately what percent" → use the percent change formula
    - "From the prior year" → the basis of comparison, 2000 total vehicle sales
    - $Percent\ Change = \frac{Focus\ -\ Basis\ of\ Comparison}{Basis\ of\ Comparison}$
    - Retrieve the necessary pieces of information from the table. Next to the ratio, place the values for each to the right. Then, simplify.
        - 2000 total vehicles = 67 or 68 thousand
        - 2001 total vehicles = 41 or 42 thousand
    - $Pct.Change = \frac{Total\ Vehicles\ 2001\ -\ Total\ Vehicles\ 2000}{Total\ Vehicles\ 2000} = \frac{42000-67000}{67000} = \frac{-25000}{67000}$
    - $Pct.Change = -37.3\%$
    - Because this value is negative, interpret it as a 37.3% decrease.
    - Choose B.
- **Behind each wrong answer choice is faulty logic**:
    - *Choice A is incorrect. It is the result of comparing 2001 to 2002 total vehicle sales. If you chose this one, you may have misread the question.*
    - *Choice B is incorrect. It is a random value you can easily eliminate via estimation. Compare 2001 to 2000 – is the total sales number in 2001 close to half of that in 2000? No, so an almost 50% change makes no sense.*

- *Choice D is incorrect. It is the result of correctly interpreting the question wording and reading the values from the chart but placing the incorrect value in the denominator of the percent change formula.*
- *Choice E is incorrect. It is a random value you can easily eliminate via estimation.*

BAR CHART 1: APPLICATIONS BY METRO AREA

A researcher reviewed government records and calculated the number of new applicants to a particular program from each of seven metropolitan areas of the United States.

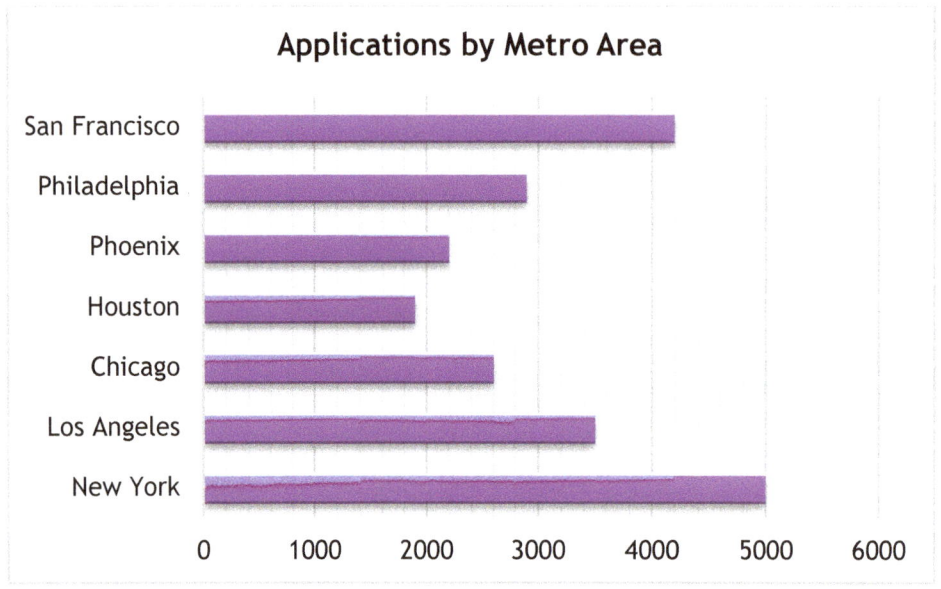

Question 1: *Refer to Bar Chart 1.* The amount by which applications from New York exceeded applications from Houston is closest to which of the following?

  a) 1900
  b) 2400
  c) 2600
  d) 3100
  e) 5000

Question 2: *Refer to Bar Chart 1.* The median number of applications came from which of these cities?

  a) Chicago
  b) Houston
  c) Los Angeles
  d) Phoenix
  e) Philadelphia

BAR CHART 1: INTERPRETING THE DATA

A researcher reviewed government records and calculated the number of new applicants to a particular program from each of seven metropolitan areas of the United States.

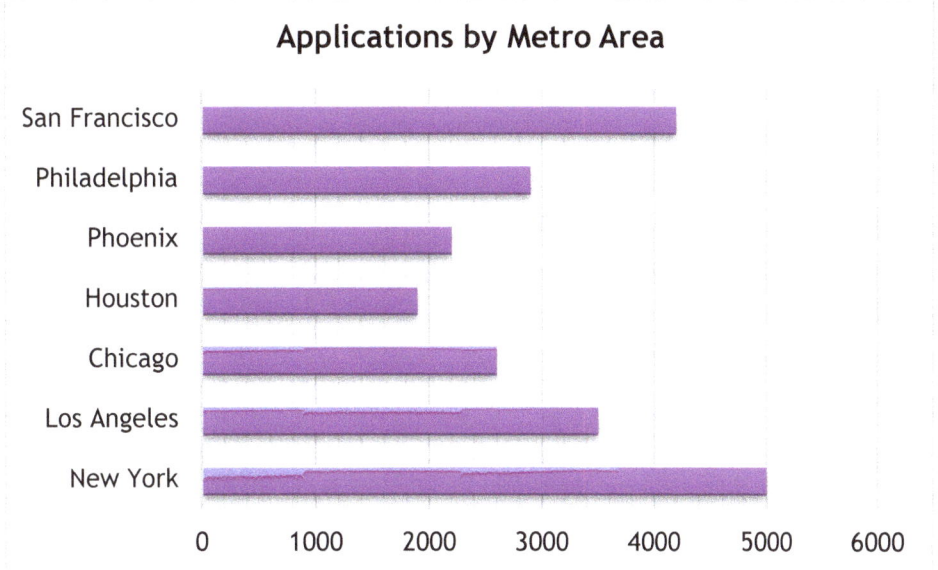

1. **Read the story** – from this, you learn that the data are about the number of applications to a program, listed by metropolitan area.
2. **Read the title** – Interest in learning various instruments
    - When – not shown in the title or the story
    - What – applications by metro area
    - Who/Where – not shown in the title, but the **Y-axis labels** tells you this includes data about 7 metro areas: San Francisco, Philadelphia, Phoenix, et al.
3. **Read the labels** and **note the units of measurement for each variable**.
    - The **X-axis** as a whole is not labeled. This is one of the few charts which will place the dependent variable on the X-axis. You can conclude that an appropriate X-axis label could be "number of applications." The major gridlines are numbered in increments of 1000. The minor gridlines are not numbered but you can quickly figure out that these are in increments of 200.
    - The **Y-axis** is labeled, showing the seven different metro areas.
4. **Read the legend** – none needed
5. **Get a sense of the data** without dwelling on the details. Focus on the big picture.
    - The largest number of applications came from New York.
    - The fewest came from Houston.
6. **Read any footnotes and/or explanations** – not applicable

Off the Charts! Data Interpretation

BAR CHART 1: ANSWERS & EXPLANATIONS

Question 1: *Refer to Bar Chart 1. The amount by which applications from New York exceeded applications from Houston is closest to which of the following?*

a) 1900
b) 2400
c) 2600
d) 3100
e) 5000

- Recognizing the type of question: **Absolute comparisons about two values**. Specifically, it's asking for the **amount by which** the applications from New York **exceeded** applications from Houston.
- Solving the question: This one is pretty straightforward.
    - "The amount by which" → you are looking for a number, not a percentage or a ratio. It's an unknown, so use a variable.
    - "Applications from New York" → refer to the chart, that's 5000.
    - "Exceeded" → this is a clue word indicating you must find the difference between two values.
    - "Applications from Houston" → refer to the chart. It's between one of the minor gridlines and the major gridline of 2000. Since each minor gridline represents 200, estimate the number for Houston as halfway between 1800 and 2000, so that's 1900.
    - $Amount = App.\,from\,New\,York - App.\,from\,Houston$
    - $Amount = 5000 - 1900 = \mathbf{3100}$
    - Choose D.
- **Behind each wrong answer choice is faulty logic**:
    - Choice A is incorrect. It is the number of applications from Houston. If you chose this one, you did part of the work, but stopped short of doing all the steps needed to arrive at the correct answer.
    - Choice B is incorrect. It is the difference between the number of applications from New York and the number of applications from Chicago. If you chose this one, your approach was correct, but you misread the chart and obtained the number for Chicago instead of Houston.
    - Choice C is incorrect. It is the number of applications from Chicago. If you chose this one, you attempted to do part of the work, but misread the chart and stopped short of doing all the steps needed to arrive at the correct answer.
    - Choice E is incorrect. It is the number of applications from New York. If you chose this one, you did part of the work, but stopped short of doing all the steps needed to arrive at the correct answer.

Question 2: *Refer to Bar Chart 1.* The median number of applications came from which of these cities?

    a) Chicago
    b) Houston
    c) Los Angeles
    d) Phoenix
    e) Philadelphia

- Recognizing the type of question: **Statistical interpretation**. Specifically, it's asking for **which city** submitted the **median number** of applications.
- Solving the question: This one is pretty straightforward.
    - Median → the middle value, when the values are arranged from smallest to largest. There are 7 cities, so the median city will be the 4th value.
    - You can count from either direction, either from smallest to largest or largest to smallest.
        - From smallest to largest: Houston, Phoenix, Chicago, <u>Philadelphia</u>
        - From largest to smallest: New York, Los Angeles, San Francisco, <u>Philadelphia</u>
    - Choose E.
- **The wrong answer choices were chosen at random for this question.**

Off the Charts! Data Interpretation

CHAPTER 7 FREQUENCY TABLES & FREQUENCY DISTRIBUTIONS

CONCEPT REVIEW

Frequency tables and **frequency distributions** are two of the next most commonly tested types of data visualizations. These tend to be more popular in fields such as market research and the social sciences and less popular in other fields.

By itself, **frequency** is a measure of how many observations fit a certain category.

- The category is typically some type of classification.
    - Think about the last time you completed a survey for a school or work. You and the other survey respondents can be classified into various categories, such as age, year in school, hair color, number of siblings, etc.
    - Some of the types above are qualitative and others are quantitative.
    - Some of the types above can be ordered whereas others have no order.
- How many observations refers to the number of observations that were recorded for a specific category.

Frequency data can be presented in either of two data visualizations. When presenting data in these forms:

- **Frequency tables**
    - The **categories** are the **independent variable** and thus should be listed on the left-most column of the frequency table. When appropriate, the categories should be ordered from least to greatest.
    - The **frequencies** (counts) for each category are the **dependent variable** and thus should be listed to the right of the corresponding category.
- **Frequency distributions**
    - The **categories** are the **independent variable** and thus should be placed along the X-axis of the frequency distribution. The categories should be ordered from least to greatest.
    - The **frequencies** (counts) for each category are the **dependent variable** and thus should be plotted along the Y-axis of the corresponding category. *A frequency distribution may look very similar to a **column chart**.*

Frequency data can be presented in either of two types:

- **Relative frequency** answers the question: how many of the responses fit this exact category?
- **Cumulative frequency** answers the question: how many of the responses fit this category and all the ones below it? Here, it is implied that the categories have already been ordered from least to greatest.

- Whenever a question involving frequency data uses **inequality** language, this wording is a clue that you have been asked to find a **cumulative frequency**.
- You will need to decipher whether to include or exclude the end point provided to you.

	Bigger	**Smaller**
Inclusive of End Point	Five <u>or more</u> <u>At least</u> 5 <u>No less than</u> 5 <u>Greater than or equal to</u> 5	Ten <u>or fewer</u> <u>At most</u> 10 <u>No more than</u> 10 <u>Less than or equal to</u> 10
Exclusive of End Point	<u>More than</u> five <u>Greater than</u> 5	<u>Fewer than</u> ten <u>Less than</u> 10

Depending upon the type of data categories you have, **cumulative frequency** may not make sense.

- If information about **nominal data** categories, such as eye color or hair color, is collected, then this information can be presented in frequency tables but only the **relative frequency** makes sense. **Cumulative frequency** makes no sense, because nominal data has no order.
- If information about **ordinal data**, **interval data**, or **ratio data** categories is collected, then this information can be presented in frequency tables using <u>either</u> **relative frequency** or **cumulative frequency**, or **both**, because each of these data types has a logical order.

Frequency data can be presented in either of these two types:
- **Absolute number** (i.e., a quantity, such as 12 or 30)
- **Relative number** (i.e., a fraction or a percentage, such as two-thirds or 20%)

Combined, there are **8 slightly different ways to create data visualizations for frequency tables & frequency distributions**. Good standardized test questions will require you to convert data obtained in one form to data in a different form.

Frequency Table	Frequency Distribution
Frequency table with relative frequencies using absolute numbers VERSION 1T	Frequency distribution with relative frequencies using absolute numbers VERSION 1D
Frequency table with relative frequencies using relative numbers VERSION 2T	Frequency distribution with relative frequencies using relative numbers VERSION 2D
Frequency table with cumulative frequencies using absolute numbers VERSION 3T	Frequency distribution with cumulative frequencies using absolute numbers VERSION 3D
Frequency table with cumulative frequencies using relative numbers VERSION 4T	Frequency distribution with cumulative frequencies using relative numbers VERSION 4D

On the following pages, the same data will be presented in each of the 8 versions listed in the table on the previous page, for the example presented at the bottom of this page.

For the example:

- Note that the categories are **mutually exclusive and collectively exhaustive** (**MECE**) (i.e., the categories do not overlap one another and cover all possible outcomes). Because of this, you can **find the total number of observations** by summing (adding up) the number of observations from any frequency data visualization with **relative frequencies** and **absolute numbers**.

- Important: If you need to convert from the **relative number** to an **absolute number** (i.e., from a percentage to a number) then you will need to know the size of the total survey population. A standardized test question may or may not give you the size of the survey population.

- Important: You need to know how to convert from a **relative frequency** to a **cumulative frequency**.

To help you understand what **mutually exclusive and collectively exhaustive** means, think about a consumer survey you may have completed recently. The age categories must be **MECE** so that all respondents fit in a single category.

Both mutually exclusive and collectively exhaustive:

What is your age?

- a) Under 18
- b) 18 to 25
- c) 26 to 35
- d) 36 to 45
- e) 46 to 55
- f) 55 and up

Mutually exclusive, but <u>not collectively exhaustive</u>:

What is your age?

- a) Under 18
- b) 18 to 25
- c) 26 to 35
- d) 36 to 45
- e) 46 to 55

A person who is 58 or 62 cannot answer the question above, because there is no answer choice which includes that person's age. For this reason, this set of answer choices is <u>not collectively exhaustive</u>.

<u>Not mutually exclusive</u>, but collectively exhaustive:

What is your age?

- a) 18 or younger
- b) 18 to 25
- c) 25 to 35
- d) 35 to 45
- e) 45 to 55
- f) 55 or older

A person who is exactly 18 could answer either A or B; a person who is exactly 25 could answer either B or C; a person who is exactly 35 could answer either C or D; and so on. For these reasons, this set of answer choices is <u>not mutually exclusive</u>.

Note: The information for each version is intentionally repetitive, to attune your attention to which elements are the same or different across the 8 versions, and what procedures needed to interpret the information are the same or different across the 8 versions.

Example: A wedding announcement website conducted a survey of 100 newly-married couples about how many children the couple planned to have.

VERSION 1T: FREQUENCY TABLE, RELATIVE FREQUENCY, ABSOLUTE NUMBERS

These data can be presented in a **frequency table** showing the **relative frequency** of responses in each category using **absolute numbers**.

Planned Number of Children per Couple	Relative Frequency (Absolute #)
0	15
1	25
2	30
3	15
4	10
5 or more	5

Basic interpretations of this version:

- 15 couples responded that they plan to have <u>zero children</u> (first row of data).
- 30 couples responded that they plan to have <u>exactly two children</u> (third row of data).
- 70 couples responded that they plan to have <u>two or fewer children</u> (sum of the first three rows of data, 15+25+30 = 70). A question asking about "a number or fewer/more" is asking you to calculate the **cumulative frequency**, inclusive.
- 5 couples responded that they plan to have <u>5 or more children</u> (last row of data), but you cannot know exactly how many. This last category thus captures all families who plan to have 5, 6, a dozen, or perhaps more children!

Statistical interpretations of this version:

- Finding the **mean** (average) number of children planned per couple: Because "5 or more" is not specific, you <u>cannot calculate the mean</u> (average) for the sample.
- Finding the **median** (middle number, when ordered) number: Find the total number of observations (N = 100), then divide by 2. Then, the median should be the average of the 50th & 51st observations. Add up the relative frequencies and stop when you've found the category which must contain the 50th & 51st observations – that's the 3rd row, or **median = 2**.
- Finding the **mode** (most common response): Look in the frequency column and find the largest number. That's 30, so the **mode = 2**.

VERSION 1D: FREQUENCY DISTRIBUTION, RELATIVE FREQUENCY, ABSOLUTE NUMBERS

The same data can also be presented in a **frequency distribution** showing the **relative frequency** of responses in each category using **absolute numbers**.

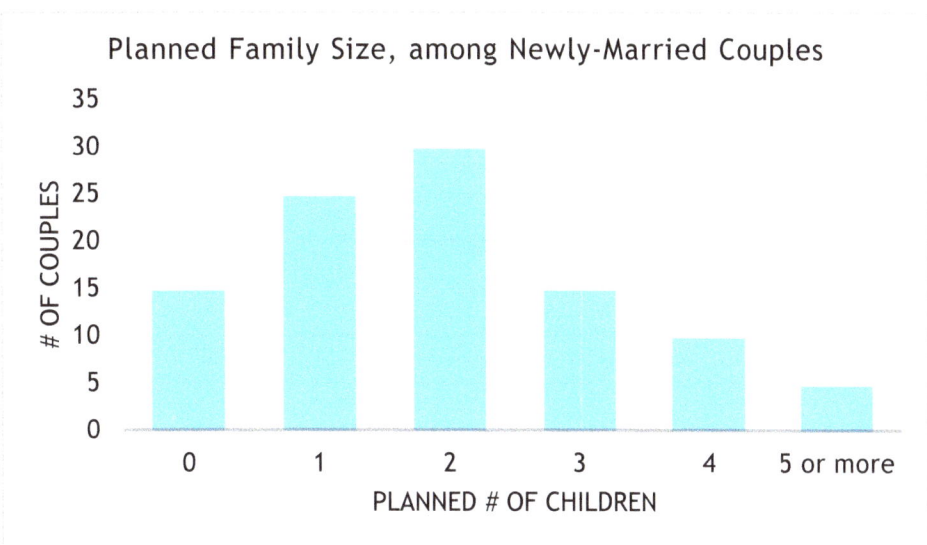

Basic interpretations of this version:
- 15 couples responded that they plan to have <u>zero children</u> (first column).
- 30 couples responded that they plan to have <u>exactly two children</u> (third column).
- 70 couples responded that they plan to have <u>two or fewer children</u>, which will be the sum of those who have exactly zero, exactly 1, or exactly 2 children. (sum of the first three columns, 15+25+30 = 70).
- 5 couples responded that they plan to have <u>5 or more children</u> (last column of data), but you cannot know exactly how many. This last category thus captures all families who plan to have 5, 6, a dozen, or perhaps more children!

Statistical interpretations of this version:
- Finding the **mean** (average) number of children planned per couple: Because "50 or more" is not specific, you <u>cannot calculate the mean</u> (average) for the sample.
- Finding the **median** (middle number, when ordered) number: Find the total number of observations (N = 100), then divide by 2. Then, the median should be the average of the 50th & 51st observations. Add up the relative frequencies and stop when you've found the category which must contain the 50th & 51st observations – that's the 3rd column, or **median = 2**.
- Finding the **mode** (most common response): Find the tallest column in the chart, then follow it down to the X-axis. That's 30, so the **mode = 2**.

VERSION 2T: FREQUENCY TABLE, CUMULATIVE FREQUENCY, ABSOLUTE NUMBERS

The same data can also be presented in a **frequency table** showing the **cumulative frequency** of responses in each category using **absolute numbers**.

Planned Number of Children per Couple	Cumulative Frequency (Absolute #)
0	15
1	40
2	70
3	85
4	95
5 or more	100

Basic interpretations of this version:

- 15 couples responded that they plan to have <u>zero children</u> (first row of data).
- 85 couples responded that they plan to have <u>three or fewer children</u> (note, the column header tells you these data are cumulative).
- To find the number of couples that plan to have <u>exactly three children</u>, you will need to subtract the cumulative frequency of couples who plan to have two or fewer children from the cumulative frequency of couples who plan to have three or fewer children.

$$Exactly\ 3\ Children\ =\ Cume.Freq.\ 3\ or\ Fewer\ -\ Cume.Freq.\ 2\ or\ Fewer$$

$$Exactly\ 3\ Children\ =\ 85 - 70$$

$$Exactly\ 3\ Children\ =\ \mathbf{15}$$

Statistical interpretations of this version:

- Finding the **mean** (average) number of children planned per couple: Because "5 or more" is not specific, you <u>cannot calculate the mean</u> (average) for the sample.
- Finding the **median** (middle number, when ordered) number: Find the total number of observations (N = 100), then divide by 2. Then, the median should be the average of the 50th & 51st observations. Find the row which contains the cumulative 50th & 51st observations – that's the 3rd row, or **median = 2**.
- Finding the **mode** (most common response): You'll need to convert from cumulative frequency back to relative frequency. Create an additional column to the right. Then, find the relative frequency for a row by subtracting the cumulative frequency for the previous row. The largest value in this new column will be your most common response. **Mode = 2**.

VERSION 2D: FREQUENCY DISTRIBUTION, CUMULATIVE FREQUENCY, ABSOLUTE NUMBERS

The same data can also be presented in a **frequency distribution** showing the **cumulative frequency** of responses in each category using **absolute numbers**.

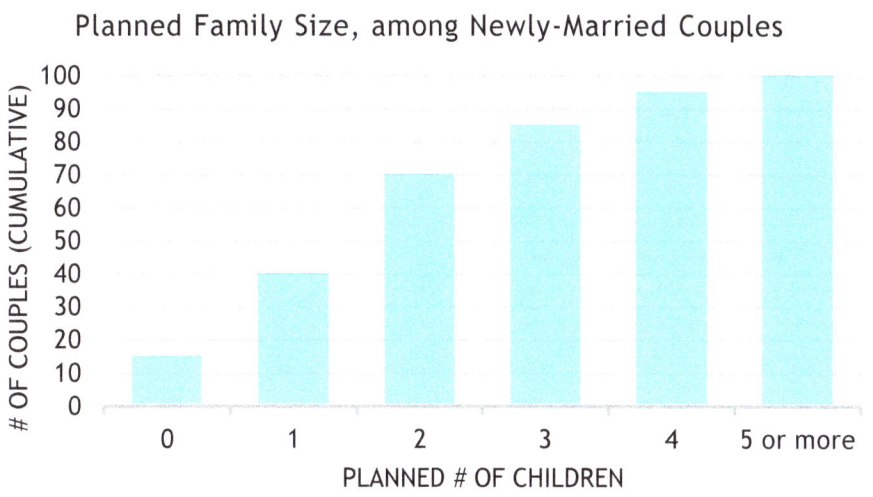

Basic interpretations of this version:

- 15 couples responded that they plan to have <u>zero children</u> (first column).
- 85 couples responded that they plan to have <u>three or fewer children</u> (note, the column label on the Y-axis tells you these data are cumulative).
- To find the number of couples that plan to have <u>exactly three children</u>, you will need to subtract the cumulative frequency of couples who plan to have two or fewer children from the cumulative frequency of couples who plan to have three or fewer children.

$$Exactly\ 3\ Children\ =\ Cume.Freq.3\ or\ Fewer\ -\ Cume.Freq.2\ or\ Fewer$$

$$Exactly\ 3\ Children\ =\ 85 - 70$$

$$Exactly\ 3\ Children\ =\ \mathbf{15}$$

Statistical interpretations of this version:

- Finding the **mean** (average) number of children planned per couple: Because "5 or more" is not specific, you <u>cannot calculate the mean</u> (average) for the sample.
- Finding the **median** (middle number, when ordered) number: Find the total number of observations (N = 100), then divide by 2. Then, the median should be the average of the 50th & 51st observations. Find the column which contains the cumulative 50th & 51st observations – that's the 3rd column, or **median = 2**.
- Finding the **mode** (most common response): You'll need to convert from cumulative frequency back to relative frequency. Create a separate set of values. Then, find the relative frequency for each column by subtracting the cumulative frequency for the previous column. The largest value in this new set of values will be your most common response. **Mode = 2**.

Off the Charts! Data Interpretation

VERSION 3T: FREQUENCY TABLE, RELATIVE FREQUENCY, RELATIVE NUMBERS

The same data can also be presented in a **frequency table** showing the **relative frequency** of responses in each category using **relative numbers** (**percentages**).

Planned Number of Children per Couple	Relative Frequency (Relative #)
0	15%
1	25%
2	30%
3	15%
4	10%
5 or more	5%

Basic interpretations of this version:

- 15% of couples responded that they plan to have <u>zero children</u> (first row of data).
- 30% of couples responded that they plan to have <u>exactly two children</u> (third row of data).
- 70% of couples responded that they plan to have <u>two or fewer children</u> (sum of the first three rows of data, 15%+25%+30% = 70%). A question asking about "a percent or fewer/more" is asking you to calculate the **cumulative frequency**.
- 5% of couples responded that they plan to have <u>5 or more children</u> (last row of data), but you cannot know exactly how many. This last category thus captures all families who plan to have 5, 6, a dozen, or perhaps more children!

Statistical interpretations of this version:

- Finding the **mean** (average) number of children planned per couple: Because "5 or more" is not specific, you <u>cannot calculate the mean</u> (average) for the sample.
- Finding the **median** (middle number, when ordered) number: The median occurs when a cumulative 50%-51% of observations is reached. Find the row which contains the cumulative 50%-51% – that's the 3rd row, or **median = 2**.
- Finding the **mode** (most common response): You'll need to convert from cumulative frequency back to relative frequency. Create an additional column to the right. Then, find the relative frequency for a row by subtracting the cumulative frequency for the previous row. The largest percentage in this new column will be your most common response. **Mode = 2**.

VERSION 3D: FREQUENCY DISTRIBUTION, RELATIVE FREQUENCY, RELATIVE NUMBERS

The same data can also be presented in a **frequency distribution** showing the **relative frequency** of responses in each category using **relative numbers** (**percentages**).

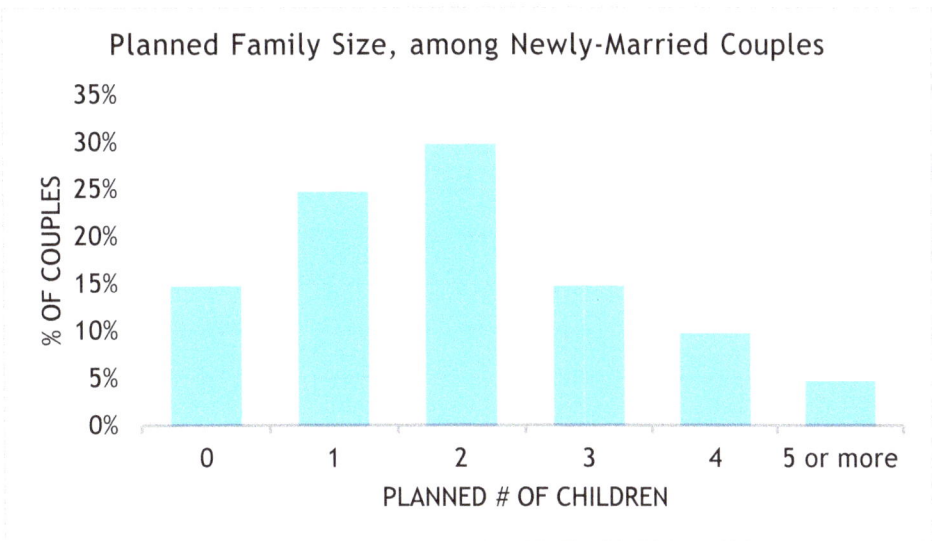

Basic interpretations of this version:

- 15% of couples responded that they plan to have zero children (first column).
- 30% of couples responded that they plan to have exactly two children (third column).
- 70% of couples responded that they plan to have two or fewer children (sum of the first three columns of data, 15%+25%+30% = 70%). A question asking about "a percent or fewer/more" is asking you to calculate the **cumulative frequency**.
- 5% of couples responded that they plan to have 5 or more children (last column of data), but you cannot know exactly how many. This last category thus captures all families who plan to have 5, 6, a dozen, or perhaps more children!

Statistical interpretations of this version:

- Finding the **mean** (average) number of children planned per couple: Because "5 or more" is not specific, you cannot calculate the mean (average) for the sample.
- Finding the **median** (middle number, when ordered) number: The median occurs when a cumulative 50%-51% of observations is reached. Find the column which contains the cumulative 50%-51% – that's the 3rd column, or **median = 2**.
- Finding the **mode** (most common response): You'll need to convert from cumulative frequency back to relative frequency. Create a separate set of values. Then, find the relative frequency for each column by subtracting the cumulative frequency for the previous column. The largest value in this new set of values will be your most common response. **Mode = 2**.

VERSION 4T: FREQUENCY TABLE, CUMULATIVE FREQUENCY, RELATIVE NUMBERS

The same data can also be presented in a **frequency table** showing the **cumulative frequency** of responses in each category using **relative numbers** (**percentages**).

Planned Number of Children per Couple	Cumulative Frequency (Relative #)
0	15%
1	40%
2	70%
3	85%
4	95%
5 or more	100%

Basic interpretations of this version:

- 15% of couples responded that they plan to have <u>zero children</u> (first row of data).
- 85% of couples responded that they plan to have <u>three or fewer children</u> (note, the column header tells you these data are cumulative).
- To find the percentage of couples that plan to have <u>exactly three children</u>, you will need to subtract the cumulative frequency of couples who plan to have two or fewer children from the cumulative frequency of couples who plan to have three or fewer children.

$$Exactly\ 3\ Children = Cume.Freq.3\ or\ Fewer - Cume.Freq.2\ or\ Fewer$$
$$Exactly\ 3\ Children = 85\% - 70\%$$
$$Exactly\ 3\ Children = \mathbf{15\%}$$

Statistical interpretations of this version:

- Finding the **mean** (average) number of children planned per couple: Because "5 or more" is not specific, you <u>cannot calculate the mean</u> (average) for the sample.

- Finding the **median** (middle number, when ordered) number: The median occurs when a cumulative 50%-51% of observations is reached. Find the row which contains the cumulative 50%-51% – that's the 3rd row, or **median = 2**.

- Finding the **mode** (most common response): You'll need to convert from cumulative frequency back to relative frequency. Create a new column of values. Then, find the relative frequency for each row by subtracting the cumulative frequency for the previous row. The largest value in this new column of values will be your most common response. **Mode = 2**.

VERSION 4D: FREQUENCY DISTRIBUTION, CUMULATIVE FREQUENCY, RELATIVE NUMBERS

The same data can also be presented in a **frequency distribution** showing the **cumulative frequency** of responses in each category using **relative numbers** (**percentages**).

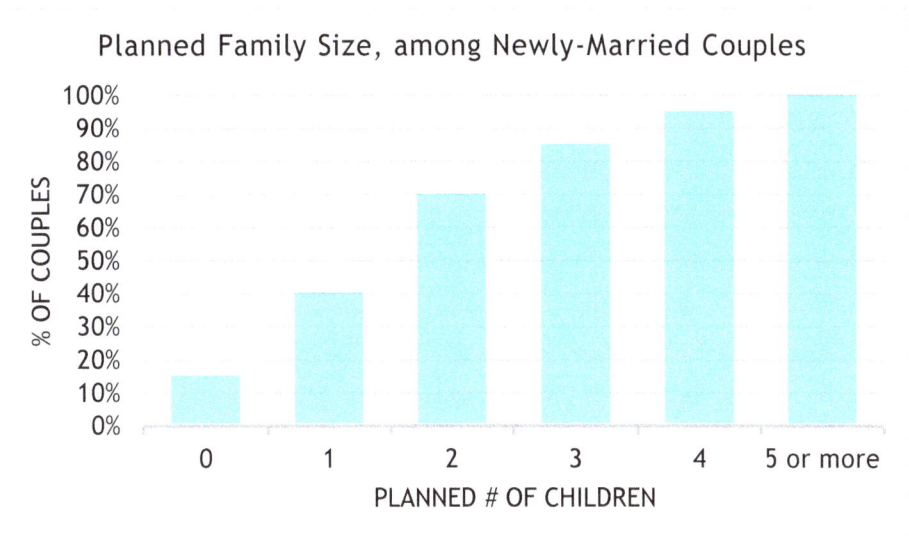

Basic interpretations of this version:

- 15% of couples responded that they plan to have <u>zero children</u> (first column).

- 85% of couples responded that they plan to have <u>three or fewer children</u> (note, the column label on the Y-axis tells you these data are cumulative).

- To find the percentage of couples that plan to have <u>exactly three children</u>, you will need to subtract the cumulative frequency of couples who plan to have two or fewer children from the cumulative frequency of couples who plan to have three or fewer children.

$$Exactly\ 3\ Children\ =\ Cume.Freq.3\ or\ Fewer\ -\ Cume.Freq.2\ or\ Fewer$$

$$Exactly\ 3\ Children\ =\ 85\%\ -\ 70\%$$

$$Exactly\ 3\ Children\ =\ \mathbf{15\%}$$

Statistical interpretations of this version:

- Finding the **mean** (average) number of children planned per couple: Because "5 or more" is not specific, you <u>cannot calculate the mean</u> (average) for the sample.

- Finding the **median** (middle number, when ordered) number: The median occurs when a cumulative 50%-51% of observations is reached. Find the column which contains the cumulative 50%-51% – that's the 3rd column, or **median = 2**.

- Finding the **mode** (most common response): You'll need to convert from cumulative frequency back to relative frequency. Create a separate set of values. Then, find the relative frequency for each column by subtracting the cumulative frequency for the previous column. The largest value in this new set of values will be your most common response. **Mode = 2**.

VERSION 5: INTEGRATED FREQUENCY DISTRIBUTION FOR SIDE-BY-SIDE COMPARISON

Spend some time working with this table, which presents, side-by-side, the relative and cumulative frequencies using both absolute numbers and relative numbers. Be sure you can convert from any given form to any of the other forms. It's the same data!

A	B	C	D	E
Planned Number of Children per Couple	Relative Frequency (Absolute #)	Cumulative Frequency (Absolute #)	Relative Frequency (Relative #)	Cumulative Frequency (Relative #)
0	15	15	15%	15%
1	25	40	25%	40%
2	30	70	30%	70%
3	15	85	15%	85%
4	10	95	10%	95%
5 or more	5	100	5%	100%

Practice working with various numbers from the table above to ensure you understand the mathematical relationship between...

Relative frequency using absolute numbers (column B) and **cumulative frequency using absolute numbers** (column C)?

$$\textit{Cume Freq. as a \#} = \textit{Sum of Relative Freq. as \# for this category \& those lesser}$$

Relative frequency using absolute numbers (column B) and **relative frequency using relative numbers** (column D)?

$$\textit{Relative Freq. as a \%} = \frac{\textit{Relative Freq. as \#}}{\textit{Sum of all Relative Freq. as \#}}$$

Cumulative frequency using absolute numbers (column C) and **cumulative frequency using relative numbers** (column E)?

$$\textit{Cume Freq. as a \%} = \frac{\textit{Cume Freq. as \#}}{\textit{Last in List Cume Freq. as \#}}$$

In short, make sure that if given <u>any</u> one of the columns B, D, C or E, and a <u>population or sample size</u>, that you could quickly calculate the values for each of the other columns.

PRACTICE SETS

FREQUENCY DISTRIBUTION 1: NUMBER OF RELATIVES

A school psychologist studying student relationships with their extended family members conducted a survey, asking each student at School Z how many total aunts and/or uncles he or she has. Their responses were recorded in the frequency distribution below.

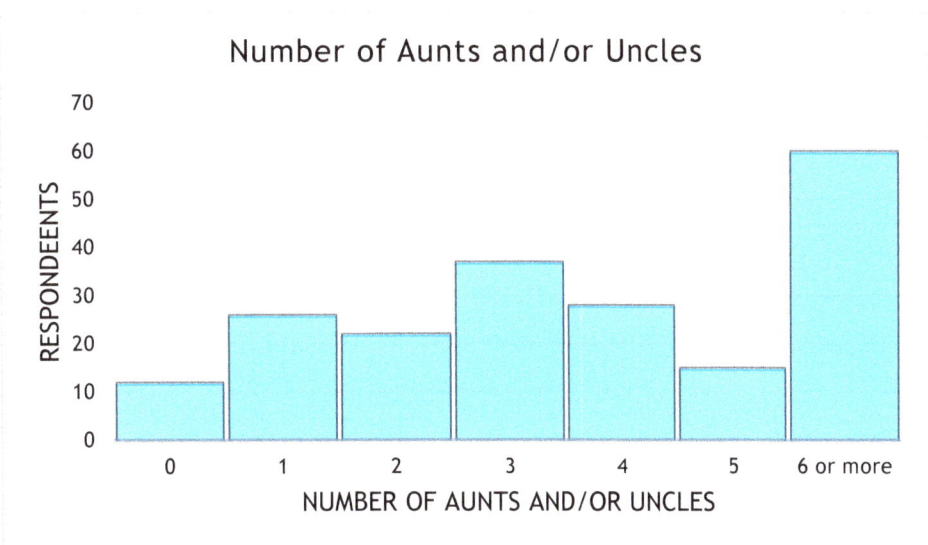

Question 1: *Refer to Frequency Distribution 1.* According to the frequency distribution, what is the median total number of aunts and/or uncles for a student at School Z?

a) 1
b) 2
c) 3
d) 4
e) 5

Question 2: *Refer to Frequency Distribution 1.* Approximately what percent of students have at least two aunts and/or uncles?

a) 11%
b) 22%
c) 81%
d) 89%
e) Cannot be determined from the information provided

Off the Charts! Data Interpretation

Question 3: *Refer to Frequency Distribution 1.* Suppose that School Z is representative of all schools in its school district. If the school district has 24,000 students overall, what is the expected number of students with exactly 3 aunts and/or uncles in this school district?

   a) 1,800
   b) 4,000
   c) 4,440
   d) 11,640
   e) Cannot be determined from the information provided

Question 4: *Refer to Frequency Distribution 1.* Approximately what percent of students have more than four aunts and/or uncles?

   a) 14%
   b) 38%
   c) 51%
   d) 86%
   e) Cannot be determined from the information provided

FREQUENCY DISTRIBUTION 1: INTERPRETING THE DATA

A school psychologist studying student relationships with their extended family members conducted a survey, asking each student at School Z how many total aunts and/or uncles he or she has. Their responses were recorded in the frequency distribution below.

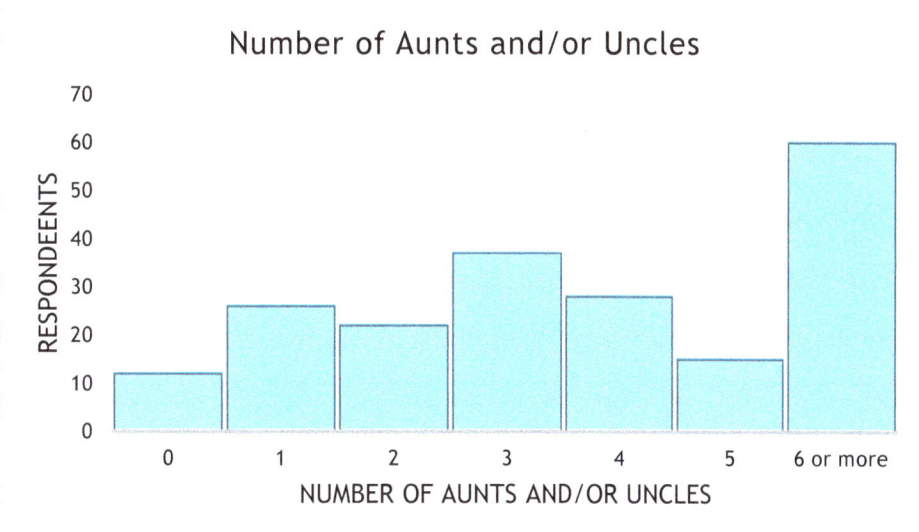

1. **Read the story** – from this, you learn that the data are about how many aunts and/or uncles various students have

2. **Read the title** – Number of Aunts and/or Uncles
    - When – not shown in the title or the story
    - What – Number of Aunts and/or Uncles
    - Who/Where – not shown in the title, but the **story** says this is for a school

3. **Read the labels** and **note the units of measurement for each variable**.
    - The **X-axis** shows how many aunts and/or uncles a student has. A "5" means a student could have 5 aunts, 5 uncles, or 5 combined.
    - The **Y-axis** is labeled, showing how many survey respondents gave that answer. *Here, because the numbers are in increments of 10 and all below 100, you must be conscientious and confirm if this axis displays a number or a percentage. Adding up the larger bars shows us this goes over 100 – so the Y-axis must represent values rather than percentages.*

4. **Read the legend** – not applicable

5. **Get a sense of the data**
    - There's an "all others" category, for "6 or more" – so you will <u>not</u> be able to calculate an average from this histogram.
    - "6 or more" was the most common response. "3" was the next most common response.

6. **Read any footnotes and/or explanations** – not applicable

Off the Charts! Data Interpretation

FREQUENCY DISTRIBUTION 1: ANSWERS & EXPLANATIONS

Question 1: *Refer to Frequency Distribution 1. According to the frequency distribution, what is the median total number of aunts and/or uncles for a student at School Z?*

   a) 1
   b) 2
   c) 3
   d) 4
   e) 5

- Recognizing the type of question: **Statistical interpretations**. It's asking you to find the **median**, or middle value.
- Finding the **median** (middle number, when ordered) number:
    - Find the total number of observations, then divide by 2. Do not concern yourself with the precision of each estimate – get "close enough" and then round off your result.
        - You an estimate half-boxes or quarter-boxes to get "close enough"
        - 12 + 25 + 22 + 36 + 28 + 15 + 60 = 198.
        - Round to a number that seems logical → 200.
        - Why? It's reasonable to assume that the underlying data had a nice, round number of observations in it. Also, your overestimations and underestimations should average out.
        - Half that = 100th / 101st observations.
    - Then, the median should be the average of the 100th & 101st observations.
    - Add up the relative frequencies and stop when you've found the category which must contain the 100th & 101st observations. It's less work to start from the right-hand side, in this case.
        - 60+15=75, +28 = 103.
        - Which column was 28 high? 4 aunts and/or uncles.
    - **Median = 4**.
    - Choose D.

Question 2: *Refer to Frequency Distribution 1. Approximately what percent of students have at least two aunts and/or uncles?*

   a) 11%
   b) 22%
   c) 81%
   d) 89%
   e) Cannot be determined from the information provided

- Recognizing the type of question: **Part-to-total comparisons.** It's asking you to determine a **percentage of** the total.
- Be careful – the wording "at least two" means you must make sure to correctly interpret this inequality.
    - "At least 2" → greater or equal to 2
    - Add up the frequencies for 2, 3, 4, 5, and 6 or more aunts and/or uncles. Note, this requires tedious effort to determine the height of 5 different bars.
    - You could alternately realize that the percentage who have "at least two" can be calculated with less effort, using (100% - percent with 0 or 1).
    - $Percent\ with\ At\ Least\ Two = 1 - Percent\ with\ Less\ than\ Two$
    - $Percent\ with\ At\ Least\ Two = 1 - \frac{Exactly\ 0 + Exactly\ 1}{Total}$
    - $Percent\ with\ At\ Least\ Two = 1 - \frac{12+25}{200} = 1 - \frac{37}{200} = \frac{163}{200} = \mathbf{81.5\%}$
- Choose C.
- **Behind each wrong answer choice is faulty logic**:
    - Choice A is incorrect because it is the percent who have <u>exactly 2</u> aunts and/or uncles. If you chose this one, you misread the question or forgot to count the cases where students had more than 2 aunts and/or uncles.
    - Choice B is incorrect. It is the height of the bar on the histogram for exactly 2 aunts and/or uncles. It fails to adjust for the size of the total, which is 200. If you chose this one, you may have worked too quickly or misinterpreted the Y-axis.
    - Choice D is incorrect. It is the result of taking choice A and subtracting it from 100%.
    - Choice E is incorrect, because you can determine an answer for this question.

Off the Charts! Data Interpretation

Question 3: *Refer to Frequency Distribution 1.* Suppose that School Z is representative of all schools in its school district. If the school district has 24,000 students overall, what is the expected number of students with exactly 3 aunts and/or uncles in this school district?

  a) 1,800
  b) 4,000
  c) 4,440
  d) 11,640
  e) Cannot be determined from the information provided

- Recognizing the type of question: **Predictions about new data points, based on trends**. For this one, you will need to **extrapolate** from the data in the chart about this school to a new value about the whole school district.
- Solving the question: Use **Words to Math** and **Words First, Math Second**.
    - "Suppose that School Z is representative of all schools in its school district." → The key phrase "representative of" means that the relative mix (percentage) of various observations in the district will be the same as the relative mix (percentage) of these observations in the school. Go to the histogram.
        - There are 37 or 38 kids with exactly 3 aunts and/or uncles.
        - There are 200 total kids.
    - "What is the expected number" → You can find an expected value by calculating the probability or proportion in one group, times the total number in the second group.
    - "of students with exactly 3 aunts and/or uncles"
    - $Exp'd \; \# \; in \; Dist \; with \; Ex. 3 = \% \; Kids \; w/ \; Exactly \; 3 \; * \; Tot. \# \; in \; District$
    - $Exp'd \; \# \; in \; Dist \; with \; Ex. 3 = \frac{\# \; Kids \; w/ \; Exactly \; 3}{Total \; in \; School} \; * \; Tot. \# \; in \; District$
    - $Exp'd \; \# \; in \; Dist \; with \; Ex. 3 = \frac{38}{200} \; * \; 24000 = \mathbf{4560}$
- Choose C.
- **Behind each wrong answer choice is faulty logic**:
    - Choice A is incorrect. This is the expected number of students in the district with exactly 5 aunts and/or uncles.
    - Choice B is incorrect. It's there as a random answer to tempt you to find a "round number" that feels satisfying.
    - Choice D is incorrect. It's the right answer to the wrong question. If you had been asked to find the number with "3 or fewer" instead of "exactly 3" then choice D would have been correct.
    - Choice E is incorrect, because you can determine an answer for this question.

Question 4: *Refer to Frequency Distribution 1. Approximately what percent of students have more than four aunts and/or uncles?*

    a) 14%
    b) 38%
    c) 51%
    d) 86%
    e) Cannot be determined from the information provided

- Recognizing the type of question: **Part-to-total comparisons**. It's asking you to determine a **percentage of** the total.
- Be careful – the wording "more than four" means you must make sure to correctly interpret this inequality.
    - "More than four" → greater than 4 = add up the categories for 5 and "6 or more"
    - Add up the frequencies for 2, 3, 4, 5, and 6 or more aunts and/or uncles to find the total number. Note, this requires tedious effort to determine the height of 5 different bars.
    - $Percent\ with\ More\ than\ 4 = Percent\ with\ 5 + Percent\ with\ 6\ or\ more$
    - $Percent\ with\ More\ than\ 4 = \frac{15 + 60}{200} = \frac{75}{200} = \mathbf{37.5\%}$
- Choose B.
- **Behind each wrong answer choice is faulty logic**:
    - Choice A is incorrect. It is the percent who have <u>exactly 4</u> aunts and/or uncles. If you chose this one, you misread the question.
    - Choice C is incorrect. It is the percent who have <u>at least 4</u> aunts and/or uncles. If you chose this one, you misinterpreted the wording of the question. Having "more than four" <u>does not include</u> having "exactly four."
    - Choice D is incorrect. It is the percent who <u>do not have exactly 4</u> aunts and/or uncles.
    - Choice E is incorrect, because you can determine an answer for this question.

FREQUENCY TABLE 2: UNIVERSITY CREDIT HOURS

A college defines student enrollment status as full-time when students are enrolled in at least 12 credit hours for the current semester. No student may take more than 18 credit hours in a single semester. The college's administrators gathered information about the number of credit hours each of its full-time students were enrolled in for the semester.

Credit Hours	# Students
12	150
13	100
14	70
15	470
16	120
17	60
18	30

Question 1: *Refer to Frequency Table 2.* What is the expected number of credit hours for a student chosen at random from the college to enroll in for a single semester?

  a) 12.0
  b) 14.3
  c) 14.6
  d) 15.0
  e) 15.2

Question 2: *Refer to Frequency Table 2.* If each student needs at least 15 credit hours per semester to stay on track to graduate within 4 years, what percent of students are not on track to graduate within 4 years?

Fill in your answer in this format: xx.x%	

FREQUENCY TABLE 2: INTERPRETING THE DATA

A college defines student enrollment status as full-time when students are enrolled in at least 12 credit hours for the current semester. No student may take more than 18 credit hours in a single semester. The college's administrators gathered information about the number of credit hours each of its full-time students were enrolled in for the semester.

Credit Hours	# Students
12	150
13	100
14	70
15	470
16	120
17	60
18	30

1. **Read the story** – from this, you learn that the data are about how many credit hours some college students enrolled in for a semester
2. **Read the title** – not applicable
    - When – not shown in the title or the story
    - What – Credit Hours & # of Students
    - Who/Where – not shown in the title, but the **story** says this is for a college
3. **Read the labels** and **note the units of measurement for each variable**.
    - The **left-most column** header the number of credit hours, from 12 to 18, in order. This is the independent variable.
    - The **right-hand column** is labeled, showing how many students fit that criteria.
4. **Read the legend** – not applicable
5. **Get a sense of the data**
    - Most students enrolled in 15 credit hours.
    - The second-most frequent category is 12 credit hours.
    - The distribution does not appear to be balanced, so the variable is not normally distributed.
6. **Read any footnotes and/or explanations** – not applicable

Off the Charts! Data Interpretation

FREQUENCY TABLE 2: ANSWERS & EXPLANATIONS

Question 1: *Refer to Frequency Table 2.* What is the expected number of credit hours for a student chosen at random from the college to enroll in for a single semester?
   a) 12.0
   b) 14.3
   c) 14.6
   d) 15.0
   e) 15.2

- Recognizing the type of question: **Statistical interpretations**. It's asking you to find the **expected value**.
- Finding the **expected value** (weighted average):
   o Steps needed to find the expected value:
      - Find the total number of observations.
      - Convert each of the absolute frequencies to relative frequencies (percentages).

      $$Relative\ Frequency = \frac{Absolute\ Frequency\ of\ Category}{Total\ Number\ of\ Observations}$$

      - Multiply each value by its relative frequency (percentage).
      - Sum these values to find the expected value.
   o You may find it easier to organize your work by <u>creating extra columns in the frequency table</u>. This organizing technique will help you avoid careless mistakes, such as forgetting to calculate one of the values or include it in your sum.

Credit Hours	# Students	% Frequency	Credit Hours * % Frequency
12	150	15.0%	1.80
13	100	10.0%	1.30
14	70	7.0%	0.98
15	470	47.0%	7.05
16	120	12.0%	1.92
17	60	6.0%	1.02
18	30	3.0%	0.54

Total Students	1000	Expected Value	14.61

- o **Expected Value = 14.61**.
- o Choose C.
- **Behind each wrong answer choice is faulty logic**:
  - o *Choice A is incorrect. If you were unfamiliar with the term **expected value**, then you may have assumed this meant the required number of credit hours.*
  - o *Choice B and E are both incorrect, but these are just extra answers designed to slow you down.*
  - o *Choice D is incorrect. It is the **mode**, or most common response, but it is not the expected value.*

Question 2: *Refer to Frequency Table 2. If each student needs at least 15 credit hours per semester to stay on track to graduate within 4 years, what percent of students are not on track to graduate within 4 years?*

Fill in your answer in this format: xx.x%	

- Recognizing the type of question: **Part-to-total comparisons**. It's asking you to determine a **percentage** of the total.
- Be careful – the wording "not on track to graduate" means you must use logic to interpret this statement and use **Words to Math**.
  - o On track to graduate → At least 15 credit hours
  - o So, "not on track to graduate" → Less than 15 credit hours.
  - o Add up the frequencies for 12, 13, and 14 credit hours. Based on the wording of this question, you would not include the students who took exactly 15 credit hours.
  - o You can either:
    - Sum the absolute frequencies, then divide by the total number of students to find the **relative frequency**.
    - Calculate each relative frequency, then add these percentages to find the **relative frequency**.
- **Answer: 32.0%**.

CHAPTER 8 HISTOGRAMS

CONCEPT REVIEW

Histograms are, essentially, a special kind of **frequency distribution**. Histograms are useful for presenting information when there is a wide variety of possible values in the data, especially for **continuous variables**.

With **continuous variables**, such as household income or business revenue, regular frequency distributions may not convey useful information because there are potentially as many distinct values as you have observations. As a result, many categories may exist with just one observation in each.

In corporate-speak, you often will "bucketize" (group) our continuous variables into math-speak "**classes**," so you have a manageable number of "buckets" or "classes."

Suppose a certain East Coast college surveyed graduates for their starting salaries and obtained the information shown in the frequency table shown below. Trying to construct a **frequency distribution** would not be useful, because each of the 18 salary amounts is unique, so you would have 18 values, each of which would have a frequency of 1.

Starting Salary	Relative Frequency
$ 41,500.00	1
$ 44,900.00	1
$ 46,250.00	1
$ 48,250.00	1
$ 51,300.00	1
$ 56,300.00	1
$ 57,500.00	1
$ 58,900.00	1
$ 59,750.00	1
$ 61,000.00	1
$ 62,000.00	1
$ 63,000.00	1
$ 67,250.00	1
$ 68,400.00	1
$ 69,000.00	1
$ 71,000.00	1
$ 73,250.00	1
$ 83,000.00	1

One way to transform data from frequency tables into histograms is to define **classes** (buckets or bands) for the variables. In this case, a reasonable **class-width** might be $10,000. You've probably seen this in consumer surveys. Now, any individual salary which falls within the defined class would be counted.

You have reduced your number of classes from 18 down to 7. Technically, two of these are empty, but you may retain them nonetheless to ensure that the resulting histogram will be **mutually exclusive**, **collectively exhaustive**. *Refer to Chapters 7 and 12 for more information about mutually exclusive and collectively exhaustive.*

As you saw with frequency distributions, the frequency information captured in histograms can capture:

- Relative frequency or cumulative frequency
- Absolute numbers or relative numbers

Starting Salary Class/Bucket	Relative Frequency (Absolute Number)	Relative Frequency (Relative Number)
Less than $40,000	0	$\frac{0}{18} = 0.0\%$
$40,000 to $49,999	4	$\frac{4}{18} = 22.2\%$
$50,000 to $59,999	5	$\frac{5}{18} = 27.8\%$
$60,000 to $69,999	6	$\frac{6}{18} = 33.3\%$
$70,000 to $79,999	2	$\frac{2}{18} = 11.1\%$
$80,000 to $89,999	1	$\frac{1}{18} = 5.6\%$
$90,000 and up	0	$\frac{0}{18} = 0.0\%$

This information is now presented in a **histogram**, with a fixed class-width of $10,000. Notice here, that this histogram shares similarities with a column chart and with a frequency distribution. One key difference – there is no gap between the columns.

When creating a **histogram**, the **classes** can be set in either of two ways:

- **Fixed-width** (i.e., each class is the same size of $10,000)
- **Fixed-number** (i.e., force *Microsoft Excel*® or other spreadsheet program to group the data into a certain number of groups. When you create quartiles or quintiles, you're essentially forcing data to fit into either 4 or 5 groups, respectively).

There is conceptual overlap between **histograms** and the statistical concepts of the **normal distribution** and **standard deviation**. *Refer to Chapter 12, which outlines strategies for word problems and review the section on Statistics.*

- You can examine the shape of the histogram to determine whether the data appear to approximate the normal distribution (bell curve) or whether the data are skewed.
- There are two types of **skewness**. **Skewness** is named for the "leg" of the data – not for the lumpy (bulky) part.

Where is the...	Positive Skew	No Significant Skew	Negative Skew
Bulk of the distribution	To the left Low end	Center	To the right High end
Leg of the distribution (outliers)	To the right High end	Equally distributed Balanced	To the left Low end
Example Diagram	Positive Skew	Symmetric (No Skew)	Negative Skew
Real-World Example	Household income There's a lower limit on income but no upper limit. A few billionaires = positive skew.	Standardized test scores These have both lower and upper limits, so extreme values do not exist.	Number of fingers on factory workers There's an upper limit on # of fingers, but you can lose them in accidents. A few amputees = negative skew.
How will mean and median relate?	Mean (average) exceeds median Those billionaires "pull up" the average.	Mean (average) roughly equal to median	Mean (average) below median Those amputees "pull down" the average.

Off the Charts! Data Interpretation

PRACTICE SETS

HISTOGRAM 1: FIRST SMARTPHONE

A market researcher conducting a study about smartphones asked 500 people "How old were you when you received your first smartphone?" and prepared a histogram.

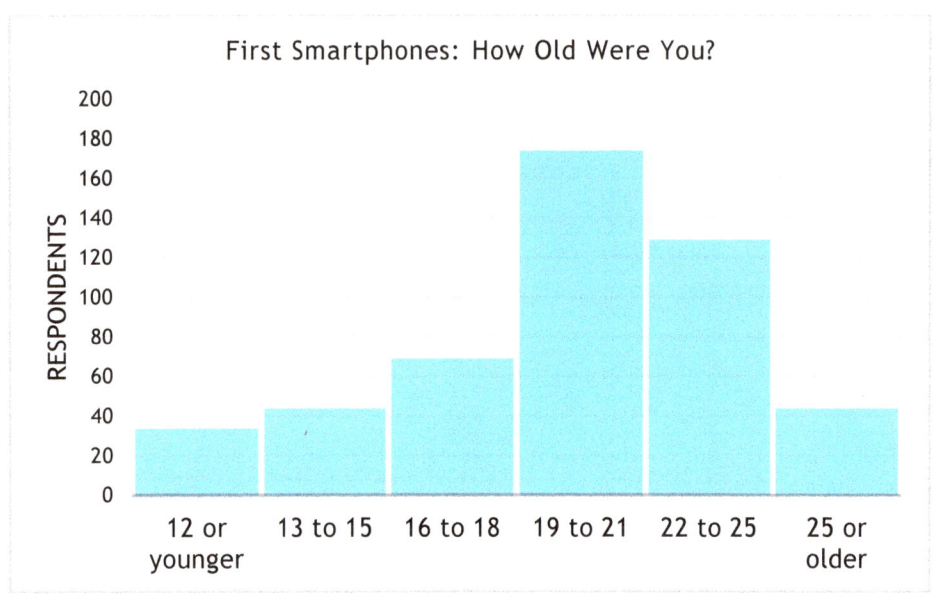

Question 1: *Refer to Histogram 1.* In which range does the median age at which the members of the survey population received their first smartphones belong?

    a) 12 or younger
    b) 13 to 15
    c) 16 to 18
    d) 19 to 21
    e) 21 to 25
    f) 25 or older

Question 2: *Refer to Histogram 1.* What percent of the survey population obtained their first smartphones after the age of 18?

    a) 9.0%
    b) 14.0%
    c) 26.0%
    d) 35.0%
    e) 70.0%

Question 3: *Refer to Histogram 1.* If a person is chosen at random from the initial survey population, what is the probability that this person was between 13 and 18 when he or she received their first smartphone?

   a) 7.0%
   b) 9.0%
   c) 14.0%
   d) 23.0%
   e) Cannot be determined from the information provided

Question 4: *Refer to Histogram 1.* If two people are chosen at random from the initial survey population, what is the probability that they were both 25 or older when they received their first smartphones?

   a) 0.0079
   b) 0.09
   c) 0.18
   d) 0.81
   e) Cannot be determined from the information provided

Question 5: *Refer to Histogram 1.* How many people were at least 13 but at most 21 when they received their first smartphones?

   a) Fewer than 100
   b) 115
   c) 245
   d) 290
   e) More than 300

HISTOGRAM 1: INTERPRETING THE DATA

A market researcher conducting a study about smartphones asked 500 people "How old were you when you received your first smartphone?" and prepared a histogram.

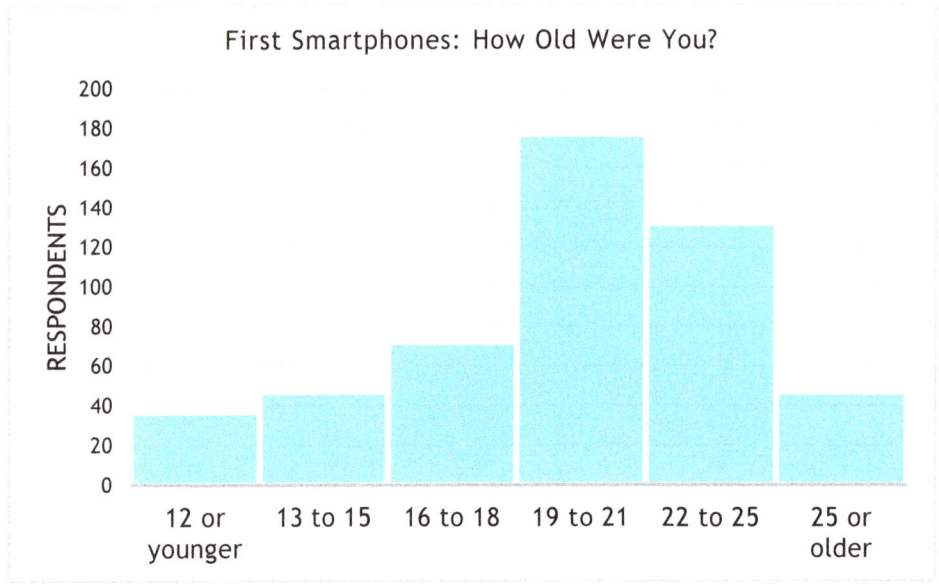

1. **Read the story** – from this, you learn that the data are about the age at which people received their first smartphones

2. **Read the title** – First Smartphones: How Old Were You
   - When – not shown in the **title** or the **story**
   - What – age at which people received their first smartphones
   - Who/Where – not shown in the **title** or the **story**

3. **Read the labels** and **note the units of measurement for each variable**.
   - The **X-axis** is not labeled, but you can logically conclude these are the age ranges from the words "or younger" and "or older". You can see that this is a histogram, because there are 6 categories of mostly equal width. The lower end ages were grouped together, and the upper end ages were grouped together, presumably to reduce the number of classes to a set which could be meaningfully understood.
   - The **Y-axis** is labeled, showing how many survey respondents gave that answer.

4. **Read the legend** – not applicable

5. **Get a sense of the data**
   - The top two categories were 19 to 21 and 22 to 25.
   - The data appear a bit skewed.

6. **Read any footnotes and/or explanations** – not applicable

HISTOGRAM 1: ANSWERS & EXPLANATIONS

Question 1: *Refer to Histogram 1.* In which range does the median age at which the members of the survey population received their first smartphones belong?

   a) 12 or younger
   b) 13 to 15
   c) 16 to 18
   d) 19 to 21
   e) 21 to 25
   f) 25 or older

- Recognizing the type of question: **Statistical interpretations**. It's asking you to find the **median**, or middle value.

- "In which range" = a clue that you do not need to calculate an exact value for the median.

- Finding the **median** (middle number, when ordered) from a histogram:
    - Find the total number of observations, then divide by 2. The **story** tells us there were 500 responses.
    - Half that = 250th / 251st observations
    - Then, the median should be the average of the 250th & 251st observations.
    - Add up the relative frequencies and stop when you've found the category which must contain the 250th & 251st observations. It's less work to start from the right-hand side, in this case.
        - 40+130=170, +175= 345.
        - Which column was 175 high? Age 19 to 21
    - **Median category = 19 to 21**.
    - Note: Even though you cannot know the <u>exact</u> median, you do know that the median will be <u>in this range</u>.
    - Choose D.

Off the Charts! Data Interpretation

Question 2: *Refer to Histogram 1.* What percent of the survey population obtained their first smartphones after the age of 18?

    a) 9.0%
    b) 14.0%
    c) 26.0%
    d) 35.0%
    e) 70.0%

- Recognizing the type of question: **Part-to-total comparisons**. It's asking you to determine a **percentage of** the total.
- Be careful – the wording "after the age of 18" means you must make sure to correctly interpret this inequality.
    - Do not include the category which includes the age of 18.
    - Add up the number of responses in the categories for **19 to 21**, **22 to 24**, and **25 or older**.
    - If you glance at the available answer choices, you see these answers are not close together – so you can safely estimate the values for each of the three categories.
    - $Percentage\ after\ 18 = \dfrac{Sum\ of\ \#\ 19\ to\ 21 + 22\ to\ 24 + 25\ or\ older}{500\ total}$
    - $Percentage\ after\ 18 = \dfrac{175 + 130 + 140}{500} = \mathbf{69\%}$
    - 69% is close to answer choice E, and not close to any others.
- Choose E.
- **Behind each wrong answer choice is faulty logic**:
    - *Choice A is incorrect. It is the percent who were age 13 to 15.*
    - *Choice B is incorrect. It is the percent who were age 16 to 18.*
    - *Choice C is incorrect. It is the percent who were age 22 to 25.*
    - *Choice D is incorrect because it is the percent who were age 19 to 21.*

Question 3: *Refer to Histogram 1. If a person is chosen at random from the initial survey population, what is the probability that this person was between 13 and 18 when he or she received their first smartphone?*

   a) 7.0%
   b) 9.0%
   c) 14.0%
   d) 23.0%
   e) Cannot be determined from the information provided

- Recognizing the type of question: **Probability of a certain outcome**.
- Solving the question: Use **Words to Math** and **Words First**, **Math Second**.
    - $Probability = \frac{\# \, Desired}{\# \, Possible}$
    - "If a person was chosen at random" → this is a single event
    - "from the initial survey population" → this is the # of <u>possible</u> outcomes
    - "what is the probability that this person" → combined with the first phrase, you must solve for the probability of a single event
    - "was between 13 and 18" → this is the # of <u>desired</u> outcomes. You must add the value of the two categories 13 to 15 and 16 to 18.
    - $Probability = \frac{\# \, in \, 13 \, to \, 15 \, + \, \# \, in \, 16 \, to \, 18}{500} = \frac{45 + 70}{500} = \frac{115}{500} = 23\%$
- Choose D.
- **Behind each wrong answer choice is faulty logic**:
    - *Choice A is incorrect. It is the probability the person was 12 or younger.*
    - *Choice B is incorrect. It is the percent who were age 13 to 15. If you chose this one, you may have calculated the percentage using 45/500, and forgotten to complete the process for the second age group, 16 to 18. This is an example of what happens when you only do <u>part of the work</u>, and then see an answer choice that matches the result of just that part.*
    - *Choice C is incorrect. It is the percent who were age 16 to 18. If you chose this one, you may have calculated the percentage using 70/500, and forgotten to complete the process for the second age group, 13 to 15. This is an example of what happens when you only do <u>part of the work</u>, and then see an answer choice that matches the result of just that part.*
    - *Choice E is incorrect because you can determine an answer.*

Question 4: *Refer to Histogram 1.* If two people are chosen at random from the initial survey population, what is the probability that they were both 25 or older when they received their first smartphones?

a) 0.0079
b) 0.09
c) 0.18
d) 0.81
e) Cannot be determined from the information provided

- Recognizing the type of question: **Probability of a certain outcome**.
- Solving the question: Use **Words to Math** and **Words First, Math Second**.
  - $Probability = \frac{\#\,Desired}{\#\,Possible}$
  - "If two people were chosen at random" → there are two events with successive probabilities
  - "from the initial survey population" → this is the # of possible outcomes
  - "what is the probability that they were both" → combined with the first phrase, you must solve for the probability of a single event
  - "25 or older" → this is the # of desired outcomes. Note: On the 2nd pick, the # Desired and the # Possible both decrease by 1.
  - $Probability = $ 1st person 25 or older AND 2nd person 25 or older
  - $Probability = \frac{\#\,25\,or\,older}{500} * \frac{\#\,25\,or\,older - 1}{500 - 1}$
  - $Probability = \frac{45}{500} * \frac{44}{499} = 0.0079$
- Choose A.
- **Behind each wrong answer choice is faulty logic**:
  - Choice B is incorrect. It is the probability of a single person in being in the group 25 or older. It fails to account for the 2nd pick.
  - Choice C is incorrect. It is the result of incorrectly interpreting the probability for the two events (you cannot add the probabilities).
  - Choice D is incorrect. It attempts to account for both events, but it makes a significant error with place value.
  - Choice E is incorrect, because the probability can be determined. Even if you were unsure of your estimate for the # of people 25 or older, choice A is pretty close and thus the best answer.

Question 5: *Refer to Histogram 1.* How many people were at least 13 but at most 21 when they received their first smartphones?

   a) Fewer than 100
   b) 115
   c) 245
   d) 290
   e) More than 300

- Recognizing the type of question: **Combining values**. It's asking you to determine a **number** for the sum of the different age groups.
- Be careful – the wording "at least 13" and "at most 21" means you have a <u>double-ended inequality</u>, so make sure to correctly interpret these phrases.
    - "At least 13" → includes all of the 13 to 15 group
    - "At most 21" → includes all of the 19 to 21 group
    - You should also <u>select all the groups in between</u>, so add the 16 to 18 group.
    - "How many" → calculate an absolute number
    - $How\ Many = (13\ to\ 15) + (16\ to\ 18) + (19\ to\ 21)$
    - $How\ Many = (45) + (70) + (175) = \mathbf{290}$
- Choose D.
- **Behind each wrong answer choice is faulty logic**:
    - Choice A and Choice E are throwaway answers.
    - Choice B is the sum of the 13 to 15 and 16 to 18 groups, but do not include the 19 to 21 group.
    - Choice C is the sum of the 16 to 18 and 19 to 21 groups, but do not include the 13 to 15 group.

Off the Charts! Data Interpretation

HISTOGRAM 2: FRESHMAN GPAS

Data regarding the GPAs of 100 freshman students were collected and presented in the following histogram:

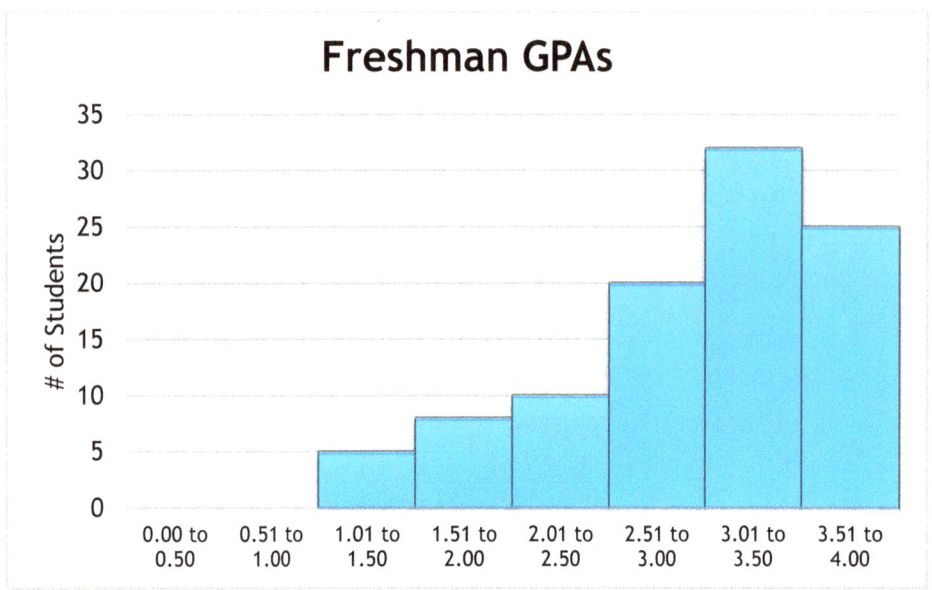

Question 1: *Refer to Histogram 2.* Based upon the information in the histogram, compare the following quantities:

Quantity A	Quantity B
The median GPA of freshman students	The range of GPAs for freshman students

a) Quantity A is greater
b) Quantity B is greater
c) Quantity A and Quantity B are equal
d) Cannot be determined

Question 2: *Refer to Histogram 2.* The ratio of the number of students with GPAs between 3.51 and 4.00 to the number of students with GPAs between 2.01 and 2.50 is approximately:

a) 2 to 5
b) 2 to 3
c) 3 to 2
d) 5 to 2

Question 3: *Refer to Histogram 2.* If two students are selected at random from this group of freshman students, what is the probability that both students will have GPAs at or below 2.0?

   a) Less than 2%
   b) 13.0%
   c) 26.0%
   d) Cannot be determined from the information given

HISTOGRAM 2: INTERPRETING THE DATA

Data regarding the GPAs of 100 freshman students were collected and presented in the following histogram:

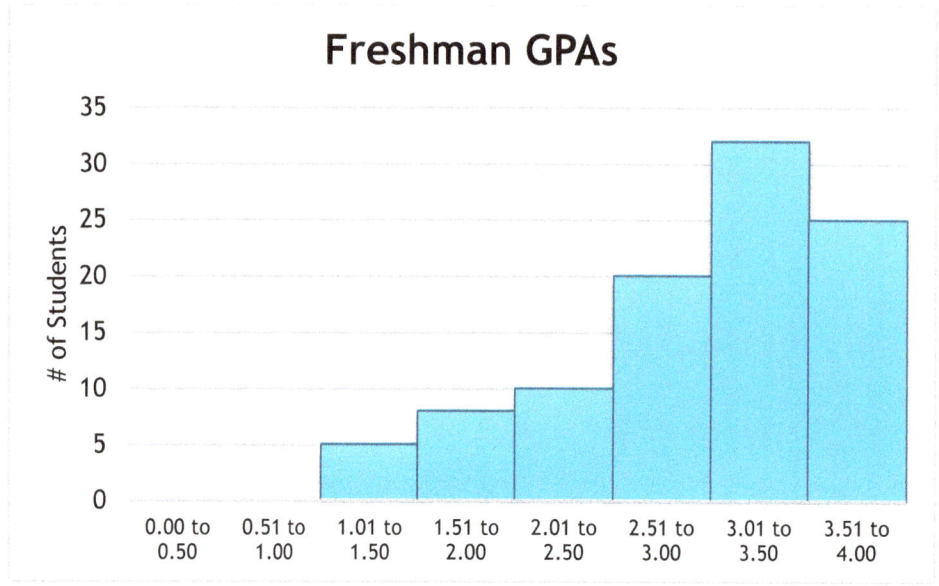

1. **Read the story** – from this, you learn that the data are about the GPAs of 100 freshman students.

2. **Read the title** – Freshman GPAs
    - When – not shown in the **title** or the **story**
    - What – freshman GPAs
    - Who/Where – not shown in the **title** or the **story**

3. **Read the labels** and **note the units of measurement for each variable**.
    - The **X-axis** is not labeled, but you can logically infer that these are GPA range. You can see that this is a histogram, not a regular column chart, because there are 8 categories of equal width. The **class width** of each column is half of a GPA point.
    - The ranges 0.00 to 0.50 and 0.51 to 1.00 do not have columns. These could have been omitted from the histogram, but the creator likely included them for a reason: to show that there are zero students with GPAs in either of these ranges.
    - The **Y-axis** is labeled, showing the number of students with GPAs in each range.

4. **Read the legend** – not applicable

5. **Get a sense of the data**
    - The most common GPA range is 3.01 to 3.50.
    - The distribution is skewed.

6. **Read any footnotes and/or explanations** – not applicable

HISTOGRAM 2: ANSWERS & EXPLANATIONS

Question 1: *Refer to Histogram 2.* Based upon the information in the histogram, compare the following quantities:

Quantity A	Quantity B
The median GPA of freshman students	The range of GPAs for freshman students

a) Quantity A is greater
b) Quantity B is greater
c) Quantity A and Quantity B are equal
d) Cannot be determined

- Recognizing the type of question: For questions which ask you to compare to quantities (typically found on the GRE®), you may need to translate one or two statements from words to math. Sometimes, the second quantity is simply a number (such as zero or 20) which requires no translation.
- Solving the question: Use **Words to Math** and **Words First**, **Math Second**.
- Evaluate Quantity A:
    o "The median GPA" → Find the median from the histogram.
    o "of freshman students" → Because the **story** and the **title** both indicate the histogram presents data about freshman students, no translation is needed.
    o Finding the **median** (middle number, when ordered) from a histogram:
    o Find the total number of observations, then divide by 2. The **story** tells us there were 100 students.
    o Half that = 50th / 51st observations
    o Then, the median should be the average of the 50th & 51st observations.
    o Add up the relative frequencies and stop when you've found the category which must contain the 50th & 51st observations. It's less work to start from the right-hand side, in this case.
        - 25 + (32 or 33) = already more than 50.
        - Which column was 32 or 33 high? GPA 3.01 to 3.50
    o **Median category = 3.01 to 3.50**
    o The <u>exact median</u> must, therefore, lie **between 3.01 and 3.50**.
- Evaluate Quantity B:
    o "The range of GPAs" → The range is the difference between the minimum and the maximum.

Off the Charts! Data Interpretation

- o "for freshman students" → Because the **story** and the **title** both indicate the histogram presents data about freshman students, no translation is needed.
  - o You do not know an exact minimum or maximum, so try to determine upper and lower limits for what the range could be.
    - Upper limit for the range (**largest possible range**):
      - The greatest possible maximum could be 4.00.
      - The least possible minimum could be 1.01.
      - Combining these, the greatest possible range would be:
      - 4.00 − 1.01 = **2.99**
    - Lower limit for the range (**smallest possible range**):
      - The least possible maximum could be 3.51.
      - The greatest possible minimum could be 1.50.
      - Combining these, the smallest possible range would be:
      - 3.51 − 1.50 = **2.01**
    - The exact range must, therefore, fall somewhere **between 2.01 and 2.99**.
- Compare your answer for Quantity A to the number in Quantity B.
  - o **Median between 3.01 and 3.50**
  - o **Range between 2.01 and 2.99**
  - o In all cases, for this question, the median will be greater than the range. Even though you do not know exact values for either the median or the range, you know that the median will always be greater than the range.
- **Quantity A is greater**.

The most common incorrect answer for this GRE-style quantity comparison question would be to choose D, that the relationship between the two quantities cannot be determined. If you chose D, you may have incorrectly assumed that because you cannot know the exact value for either the median or the range, that you could not evaluate which is greater. In this case, you can!

Question 2: *Refer to Histogram 2. The ratio of the number of students with GPAs between 3.51 and 4.00 to the number of students with GPAs between 2.01 and 2.50 is approximately:*

   a) 2 to 5
   b) 2 to 3
   c) 3 to 2
   d) 5 to 2

- Recognizing the type of question: **Part-to-part comparison**. Specifically, it's asking for the **ratio** of info about one group (a part) to the same info about a different group (a part).

- Solving the question: Use **Words to Math** and **Words First**, **Math Second**.
    - "The ratio of..." → Set up a fraction
    - "the number of students with GPAs between 3.51 and 4.00" → This comes first, so it goes in the numerator of your fraction
    - "to the number of students with GPAs between 2.01 and 2.50" → This comes second, so it goes in the denominator of your fraction
    - Retrieve the necessary pieces of information from the table. Next to the ratio, place the values for each to the right. Then, simplify.
    - $\frac{\text{\# of Students with GPAs between 3.51 and 4.00}}{\text{\# of Students with GPAs between 2.01 and 2.50}} = \frac{25}{10} = \frac{5}{2} = 5\ to\ 2$
    - Choose D.

- **Behind each wrong answer choice is faulty logic**:
    - Choice A is incorrect because it is the reciprocal of the correct answer. Be cautious with ratios so that you put the right part in the right place!
    - Choice B is incorrect for two reasons. It is the reciprocal of answer choice C. See that explanation below.
    - Choice C is incorrect because it is the right answer to the wrong question. Here, it's the correct ratio for the number of students with GPAs between 3.01 and 3.50 to the number of students with GPAs between 2.51 and 3.00. If you chose this one, you may have either misread the question or misread the histogram.

Question 3: *Refer to Histogram 2. If two students are selected at random from this group of freshman students, what is the probability that both students will have GPAs at or below 2.0?*

  a) Less than 2%
  b) 13.0%
  c) 26.0%
  d) Cannot be determined from the information given

- Recognizing the type of question: **Probability**. In this case, you are asked to find a **two-event probability**.
- Solving the question: Use **Words to Math** and **Words First, Math Second**.
    - $Probability = \frac{\# \, Desired}{\# \, Possible}$
    - $Probability \, of \, Two \, Events = P(1st \, event) \, AND \, P(2nd \, event)$
    - "If two students are selected at random" → There are two events, the selection of the first student and the selection of the second student.
    - "from this group of freshman students" → This is the # of possible outcomes
    - "what is the probability that both students" → Anticipate that the criteria will come next. Both the first student and the second student will need to meet the stated criteria. Remember that with probability, the word AND should prompt you to multiply.
    - "will have GPAs at or below 2.0" → this is the # of desired outcomes. You must add the number of observations in all categories at or below 2.0.
    - From the histogram, there are 5 in the 1.01 to 1.50 group and approximately 7 or 8 in the 1.51 to 2.00 group. There is a total of 12 or 13 students with GPAs "at or below 2.0."
    - Does precision matter for this question? No! Take a quick glance at the answer choices. These answers are not that close to each other, so do not worry about obtaining precise values from the histogram.
    - $Probability = P(1st \, event) * P(2nd \, event)$
    - $Probability = \frac{13}{100} * \frac{12}{99} = \mathbf{0.0158 \, or \, 1.6\%}$
- Choose A.
- **Behind each wrong answer choice is faulty logic**:
    - *Choice B is incorrect. It is the probability that a single student had a GPA at or below 2.0.*
    - *Choice C is incorrect. It is the result of adding the probability of each student having a GPA at or below 2.0. Recall that when you are working with the probability of multiple events, the word AND means multiply.*
    - *Choice D is incorrect because you can determine an answer.*

CHAPTER 9 BOX-AND-WHISKER PLOTS

CONCEPT REVIEW

Box-and-whisker plots are a type of data visualization which capture the essential pieces of information about the statistical distribution of a given data set. A **box-and-whisker plot** will visually convey the following:

- **Minimum value** and **maximum value**
- **Median value**
- The **range** of the entire data set
- The **range** of each **quartile**, which is calculated by subtracting the value at the lower end from the value at the higher end of the quartile.
- Whether the **distribution** appears to be **symmetric** or **skewed**
- The **mean** (average) may or may not be depicted on a box-and-whisker plot. It is typically denoted with a dot, X, or diamond-shaped marker.
- Any **outliers** which would be indicated by dots on the plot beyond the whiskers.

A **box-and-whisker plot** is often oriented horizontally and may or may not include an X-axis. Examine the diagram below, then look at the table to understand how to interpret the statistical information conveyed.

Statistical Information	Value	How You Find It
Maximum Value	350	The maximum is represented by the far-right whisker
Minimum Value	10	The minimum is represented by the far-left whisker
Median Value	75	The median is represented by the line in the middle of the box
1st quartile range	10 to 40	The 1st quartile range is represented by the distance from the minimum to the lower end of the box
2nd quartile range	40 to 75	The 2nd quartile range is represented by the distance from the lower end of the box to the line in the middle of the box

Off the Charts! Data Interpretation

Statistical Information	Value	How You Find It
3rd quartile range	75 to 120	The 3rd quartile range is represented by the distance from the line in the middle of the box to the upper end of the box
4th quartile range	120 to 350	The 4th quartile range is represented by the distance from the upper end of the box to the maximum
Middle 50% range	40 to 120	The middle 50% range is represented by the box between the two whiskers
Shape of the data distribution	Positively skewed	Visually inspect the diagram. • If it appears to be symmetric, the underlying data set would be balanced (not skewed) and possibly a normal distribution. • If it does not appear to be symmetric, then determine the direction of the skewness. Compare the 1st quartile range and the 4th quartile range. Which has a wider range? • Negative skewness = if the 1st quartile range is larger (line is longer) • Positive skewness = if the 4th quartile range is larger (line is longer)

Box plots are, essentially, a more complex variation of **box-and-whisker diagrams**.

- A **box plot** typically has a vertical orientation, whereas a **box-and-whisker plot** typically has a horizontal orientation.
- A **box plot** may place several of these vertically-oriented **box-and-whisker plots** side-by-side to facilitate comparison of related data sets.

Thus, a **box plot** conveys the same information as a box-and-whisker plot, but allows you to take your analysis a step further and compare the statistical information about one data set with that of another data set:

- **Minimum value** and **maximum value**
- **Median value**
- The **range** of the entire data set
- The **range** of each **quartile**, which is calculated by subtracting the value at the lower end from the value at the higher end of the quartile
- Whether the **distribution** appears to be **symmetric** or **skewed**, as well as the direction of the skewness
- Any **outliers** which would be indicated by dots on the plot beyond the whiskers.

Examine the **box plot** on the next page, then look at the table to understand how to interpret the statistical information conveyed. In this variation, you'll see the **box plot** contains information about 3 groups, which can be identified using the **legend**. The **major gridlines** are marked in increments of 50. Because there are 4 **minor gridlines** to the next major gridline (for 5 lines in total), then you know each minor gridline represents an increment of 10.

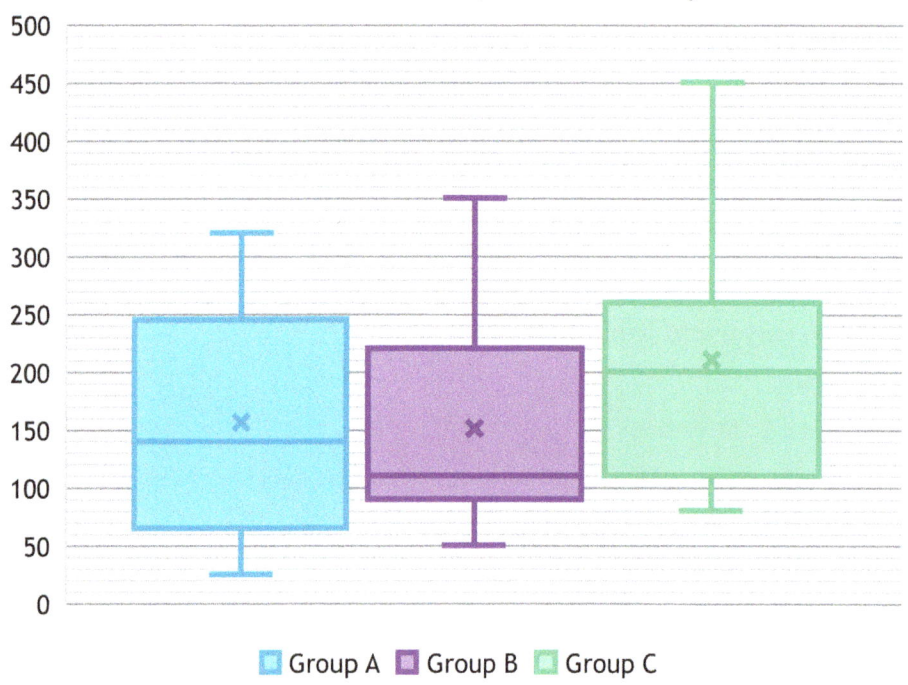

Statistical Information	Value Gp A	Value Gp B	Value Gp C	How You Find It
Maximum Value	320	350	450	The maximum is represented by the <u>upper-most</u> whisker
Minimum Value	25	50	80	The minimum is represented by the <u>lower-most</u> whisker
Median Value	140	110	200	The median is represented by the horizontal line in the middle of the box
Mean Value	Est. 150	Est. 150	Est. 210	The mean is represented by the special marker, in this case an X, but in other cases a diamond or a circle may be used. The marker for the mean may also be omitted.
1st quartile range	25 to 60	50 to 90	80 to 110	The 1st quartile range is represented by the distance from the minimum to the lower end of the box
2nd quartile range	60 to 140	90 to 110	110 to 200	The 2nd quartile range is represented by the distance from the lower end of the box to the line in the middle of the box

Statistical Information	Value Gp A	Value Gp B	Value Gp C	How You Find It
3rd quartile range	140 to 250	110 to 220	200 to 260	The 3rd quartile range is represented by the distance from the line in the middle of the box to the upper end of the box
4th quartile range	250 to 320	220 to 350	260 to 450	The 4th quartile range is represented by the distance from the upper end of the box to the maximum
Middle 50% range	60 to 250	90 to 220	110 to 260	The middle 50% range is represented by the box between the two whiskers
Shape of the data distribution	Slight pos. skew	Pos. Skew	Pos. Skew	Visually inspect the diagram. • If it appears to be symmetric, the underlying data set would be balanced (not skewed) and possibly a normal distribution. • If it does not appear to be symmetric, then determine the direction of the skewness. Compare the 1st quartile range and the 4th quartile range. Which has a wider range? • Negative skewness = if the 1st quartile range is larger (line is longer) • Positive skewness = if the 4th quartile range is larger (line is longer)

Off the Charts! Data Interpretation

PRACTICE SETS

BOX-AND-WHISKER PLOT 1: LANGUAGE DEVELOPMENT

A researcher who is interested in the development of language among children has gathered data regarding the number of words in the vocabulary of 4-year-old children.

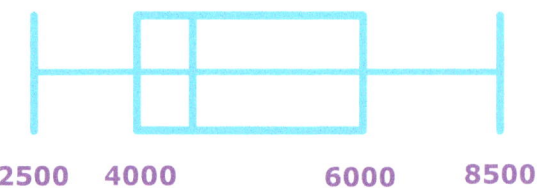

2500   4000   6000   8500

Question 1: *Refer to Box-and-Whisker Plot 1.* Which of the following statements must be true, based upon the information shown? Choose all that apply.

a) The median number of words in the vocabulary of 4-year-olds is approximately 4,500 words.
b) The average number of words in the vocabulary of 4-year-olds is approximately 4,500 words.
c) A 4-year-old child who knows more than 5,000 words must be in the top 25% of the peer group.

Question 2: *Refer to Box-and-Whisker Plot 1.* Based upon the box-and-whisker plot, it can reasonably be concluded that the shape of the distribution of the data set:

a) Approximates the normal distribution
b) Is positively skewed
c) Is negatively skewed

BOX-AND-WHISKER PLOT 1: INTERPRETING THE DATA

A researcher who is interested in the development of language among children has gathered data regarding the number of words in the vocabulary of 4-year-old children.

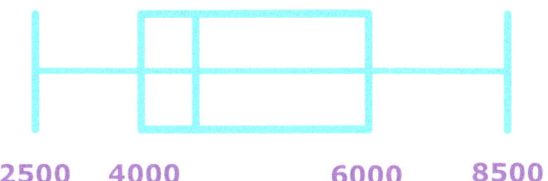

1. **Read the story** – from this, you learn that the data gathered are about the number of words spoken by 4-year-old children.

2. **Read the title** – not applicable
    - When – not shown in the **story** or the **title**
    - What – number of words spoken by 4-year-old children, based on the **story**
    - Who/Where – not shown in the **story** or the **title**

3. **Read the axis labels** and **note the units of measurement for each variable**.
    - No axis is used – but the various data points are labeled.

4. **Read the legend** – not applicable

5. **Get a sense of the data** without dwelling on the details. Focus on the big picture.
    - The **range** is from 2500 to 8500 words.
    - The **middle 50% range** is from 4000 to 6000 words.
    - The **median** is not labeled, but it is closer to 4000 than 6000.

6. **Read any footnotes and/or explanations** – not applicable

Off the Charts! Data Interpretation

BOX-AND-WHISKER PLOT 1: ANSWERS & EXPLANATIONS

Question 1: *Refer to Box-and-Whisker Plot 1.* Which of the following statements must be true, based upon the information shown? Choose all that apply.

a) The median number of words in the vocabulary of 4-year-olds is approximately 4,500 words.
b) The average number of words in the vocabulary of 4-year-olds is approximately 4,500 words.
c) A 4-year-old child who knows more than 5,000 words must be in the top 25% of the peer group.

- Recognizing the type of question: Each statement may require a different approach, so you'll need to use the same general process but different specific steps. With box-and-whisker plots, you can expect nearly all of the questions will involve **statistical interpretation**.
- Solving the question: Use **Words to Math** and **Words First, Math Second**.
- Evaluate Statement A:
    - "The median number" → Find the middle value, shown by the center line.
    - "Approximately" → You can estimate, so you do not need to be precise.
    - There is neither an X-axis nor a data label on the center line, so you must attempt to visually divide the box into equal sections. If you divide the box in half, then half again, you'll see that the center line is approximately one-fourth of the distance from 4,000 to 6,000.
    - Thus, 4,500 is a reasonable estimate for the median.
    - **Statement A is true**.
- Evaluate Statement B:
    - "The average number" → Find the average or mean value
    - "Approximately" → You can estimate, so you do not need to be precise.
    - There is no marker for the mean on this box-and-whisker plot, so you would need to know each of the underlying values to determine an average.
    - The trick is to avoid mixing up the words and meanings of **mean** and **median**.
    - **Statement B is false**.
- Evaluate Statement C:
    - "more than 5,000 words" → There is neither an X-axis nor a data label on the center line, so you must attempt to visually divide the box into equal sections. You can divide the box in half to place an observation of 5,000 words onto the box-and-whisker plot.
    - "must be" → always true, with no exceptions
    - "in the top 25% of the peer group" → in other words, in the 4th quartile.

- o Because 5,000 words falls between the median and the end of the box, you know that a 4-year-old child with a vocabulary of "more than 5,000" words will fall somewhere between the middle of the 3rd quartile and the end of the 4th quartile.
- o The <u>actual value</u> could be 5,700 words – and the statement would be false. The <u>actual value</u> could also be 7,300 words – and the statement would be true.
- o This <u>could be</u> true, but it is <u>not necessarily</u> true.
- o **Statement C is false**.
- **You must choose A, but not B or C, to get the point for the question**.

Question 2: *Refer to Box-and-Whisker Plot 1.* Based upon the box-and-whisker plot, it can reasonably be concluded that the shape of the distribution of the data set:

   a) Approximates the normal distribution
   b) Is positively skewed
   c) Is negatively skewed

- Recognizing the type of question: Because the question asks about the shape of the distribution, this is a **statistical interpretation** question.
- Evaluate Statement A:
    - Recall that the normal distribution should be symmetrical around the mean.
    - Is this box-and-whisker plot symmetrical? No.
    - **Statement A is false**.
- Evaluate Statements B & C together, because these are opposites.
    - Recall that the skewness of a distribution is named for the "leg" of the data – not for the lumpy (bulky) part.
    - Which whisker extends further, beyond the box? If you need to be precise, calculate the range of the 1st and 4th quartiles.
        - If it's the left whisker, then there is negative skewness. The range of Q1 is from 2,500 to 4,000 = 1,500.
        - If it's the right whisker, then there is positive skewness. The range of Q4 is from 6,000 to 8,500 = 2,500.
        - The whisker on the right extends further (the range of Q4 is greater than the range of Q1).
        - Thus, this distribution would exhibit positive skewness.
    - **Statement B is true**.
    - **Statement C is false**.
- **You must choose B, but not A or C, to get the point for the question**.

BOX PLOT 2: ATHLETE STATURE

A website which collects statistics about athletes participating in multiple sports gathered information about the body weight of football players who played the linebacker position at three levels of increasing competition: high school, college, and professional.

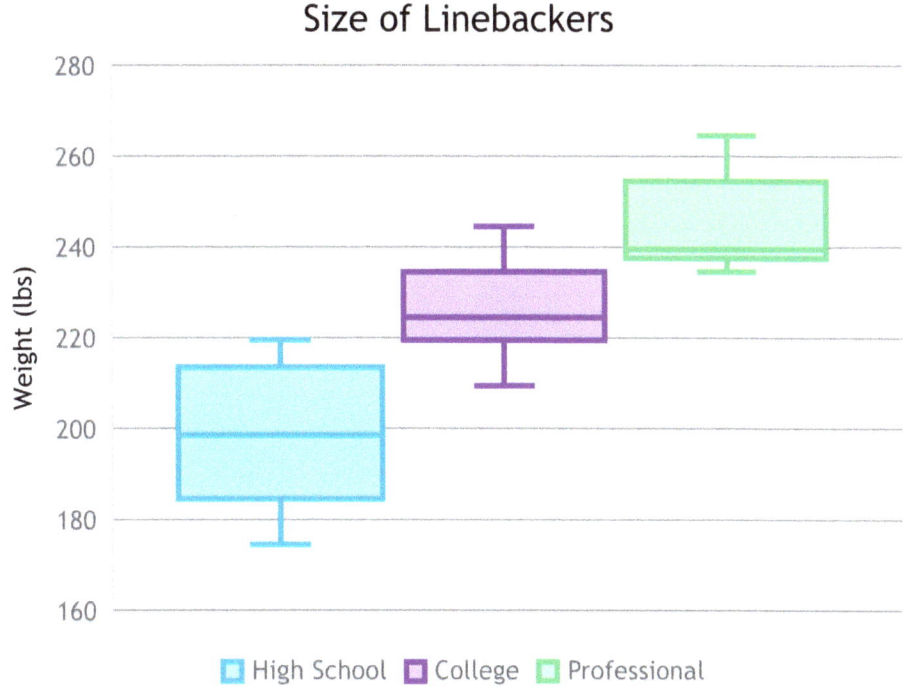

Question 1: *Refer to Box Plot 2.* Which of the following statements are true, based upon the information shown? Choose all that apply.

   a) The median weight of a high school linebacker is approximately 200 pounds.
   b) The maximum weight of professional linebackers is approximately 255 pounds.
   c) The range of weights of high school linebackers is at least 40 pounds.

Question 2: *Refer to Box Plot 2.* The range of the 4th quartile of professional linebackers is approximately how much greater than the range of the 1st quartile of professional linebackers?

   a) 0
   b) 6
   c) 10

Off the Charts! Data Interpretation

BOX PLOT 2: INTERPRETING THE DATA

A website which collects statistics about athletes participating in multiple sports gathered information about the body weight of football players who played the linebacker position at three levels of increasing competition: high school, college, and professional.

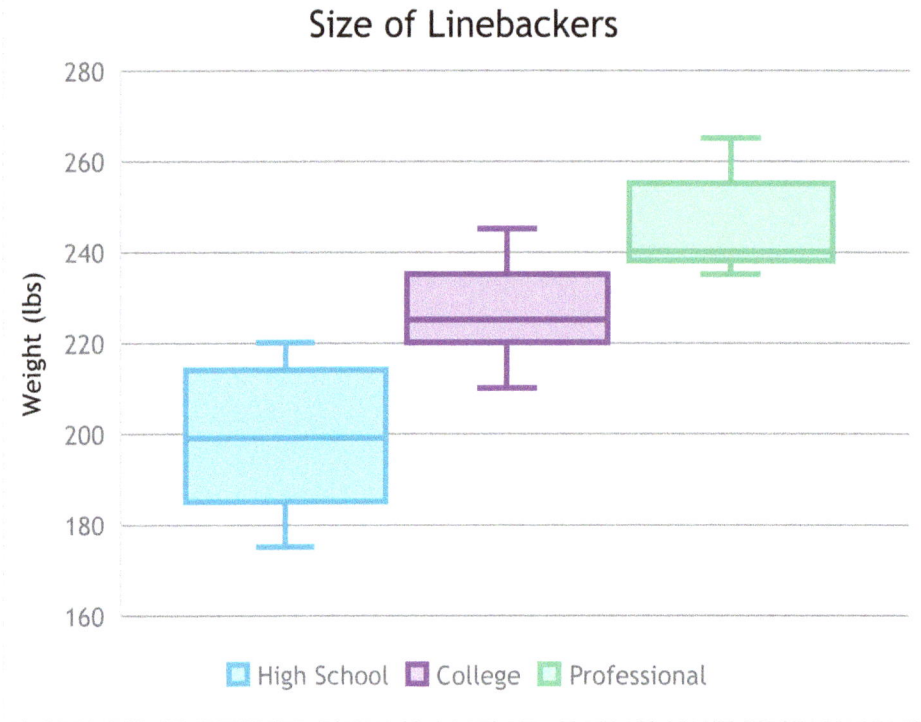

1. **Read the story** – from this, you learn that the data gathered is about the body weight of athletes at three levels of competition.

2. **Read the title** – Linebacker position
    - When – not shown in the **story** or the **title**
    - What – body weight of linebackers in high school, college, and professional. Combined, the **story**, **legend**, and **Y-axis label** say this.
    - Who/Where – not shown in the **story** or the **title**

3. **Read the axis labels** and **note the units of measurement for each variable**.
    - The Y-axis denotes weight in pounds. If you look closely, you can see that the Y-axis does not start at the origin (zero).

4. **Read the legend**
    - Blue box plot on the left = high school players
    - Purple box plot in the middle = college players
    - Green box plot on the right = professional players

5. **Get a sense of the data** without dwelling on the details. Focus on the big picture.

Off the Charts! Data Interpretation

- The **median** body weight of players increases at higher levels of competition.
- The **range** is smaller at higher levels of competition.
- The **maximum** weight of a high school player is less than the **minimum** weight of a professional player. (The blue and green box plots do not overlap at all).

6. **Read any footnotes and/or explanations** – not applicable

BOX PLOT 2: ANSWERS & EXPLANATIONS

Question 1: *Refer to Box Plot 2.* Which of the following statements are true, based upon the information shown? Choose all that apply.
   a) The median weight of a high school linebacker is approximately 200 pounds.
   b) The maximum weight of professional linebackers is approximately 255 pounds.
   c) The range of weights of high school linebackers is at least 40 pounds.

- Recognizing the type of question: Each statement may require a different approach, so you'll need to use the same general process but different specific steps.
- Solving the question: Use **Words to Math** and **Words First, Math Second**.
- Evaluate Statement A:
   - "The median weight" → Find the middle value, shown by the center line.
   - "of a high school linebacker" → Look at the legend. You'll need to use the blue box plot on the left.
   - "is approximately 200 pounds" → The word "approximately" suggests that you can estimate the value. Do not waste effort trying to get an exact value.
   - The middle line in the blue box plot is either just at or just below 200, so **Statement A is true**.
- Evaluate Statement B:
   - "The maximum weight" → Find the largest value, by looking at the whisker on the top
   - "of professional linebackers" → Look at the legend. You'll need to use the green box plot on the right.
   - "is approximately 255 pounds" → The word "approximately" suggests that you can estimate the value.
   - Because the maximum is above the major gridline for 260 pounds, you do not need to try to interpret the precise value for the maximum. Anything larger than 260 is also larger than 255.
   - **Statement B is false**.
- Evaluate Statement C:
   - "The range of weights" → Find the difference between the maximum & the minimum
   - "of high school linebackers" → Look at the legend. You'll need to use the blue box plot on the left.
   - "is at least 40 pounds" → The phrase "at least" suggests that you can estimate the value.
      - Since the major gridlines are marked in increments of 20 pounds, check to see if the box plot crosses at least 2 major gridlines.

- - - Alternatively, you can try to estimate both the maximum and the minimum, then subtract.
  - Because the range crosses two full gridlines, plus a little extra, you know that the range is greater than 40 pounds.
  - **Statement C is true**.
- **You must choose both A and C, but not B, to get the point for the question**.

Question 2: *Refer to Box Plot 2.* The range of the 4th quartile of professional linebackers is approximately how much greater than the range of the 1st quartile of professional linebackers?

    a) 0
    b) 6
    c) 10

- Recognizing the type of question: **Statistical interpretation** and **absolute comparisons about two values**.
- Solving the question: Use **Words to Math** and **Words First, Math Second**.
    - "The range of the 4th quartile" → you will need the find the distance from the top of the box to the top whisker.
        - Estimate between the major gridlines. Each increment of 20 could be cut in half, then half again, so each quarter box is roughly 5 pounds.
        - Top whisker = approx. 265
        - Top of box = approx. 255
    - "of professional linebackers" → look at the legend. You'll need to use the green box plot on the right.
    - "is approximately how much greater than" → you will need to subtract.
    - "the range of the 1st quartile → you will need the find the distance from the bottom whisker to the bottom of the box.
        - Estimate between the major gridlines. Each increment of 20 could be cut in half, then half again, so each quarter box is roughly 5 pounds.
        - Bottom whisker = approx. 235
        - Bottom of box = not quite on the line at 240, so approx. 237 or 238
- *Interquartile Range of 4th* − *Interquartile Range of 1st* =
- $(265 - 255) - (238 - 235) = 10 - 3 = 7$
- **Choose B.**
- **Behind each wrong answer choice is faulty logic**:
    - Choice A is incorrect. It uses information from the box plot about college-level players.
    - Choice C is incorrect. This is the range of the 4th quartile about professional players. If you chose this answer, you may have done the first step, but forgotten to perform the second step.

CHAPTER 10 SCATTERPLOTS

CONCEPT REVIEW

A **scatterplot** is a type of data visualization which shows the value of each data point in the data set in the XY-coordinate plane

Scatterplots are often used to visualize the relationship between two variables which are correlated (i.e., the two variables tend to vary in relation to one another) but do not necessarily have a cause-and-effect relationship.

A **trendline** may be added to the scatterplot to express the mathematical relationship between the two variables. A scatterplot can have:

- A **linear trendline** – this is a straight line, with the **linear trendline equation** expressed in the **standard slope-intercept form** of:

$$Y = MX + B$$

Using words, this same slope-intercept form can be understood, retained, and applied to various situations described in word problems using:

$$Ending = Rate * Quantity + Beginning$$

- An **exponential trendline** – this is a curved line, with the **exponential trendline equation** expressed in the **standard exponential form** of:

$$Y = A * K^X$$

Using words, this same exponential form can be understood, retained, and applied to various situations described in word problems using:

$$Ending = Beginning * (1 + Change)^{Time}$$

For a visual depiction of the various types of trendlines, refer to Chapter 2, in the section Describing Trends in Data.

There is a bit of an art to choosing the right type of trendline for your scatterplot. It requires a combination of visual inspection to get a sense of whether the trend is linear or curved and using software to calculate the trendline equation based on whichever type you select as the best-fit for the data.

On standardized tests, however, you will not need to go through this iterative process – instead, you should understand such basics about a trendline as:

- Is it linear or curved?
- Is the general trend increasing or decreasing?
- What is the Y-intercept? If needed, extend a linear trendline toward the Y-axis to estimate the Y-intercept value.
- To predict values not marked on the graph using the trendline, extend the trendline in the appropriate direction, taking care to maintain a consistent width of the X-axis increments and consistent height of the Y-axis increments. You'll need to do one of these two things:

- o **Extrapolating**: Extend the trendline <u>beyond</u> those on the graph
- o **Interpolating**: Estimate <u>between</u> two data points on the graph
- How well does the trendline explain the **variation** in the data? In other words, what is the <u>strength of the relationship</u> between the two variables?
    - o If you are asked to do this, you look at the R-squared value associated with the trendline.
    - o R-squared values are decimals which vary between 0.00 and 1.00.
    - o *Look at the guidelines in the following table to understand how to interpret R-squared values, if you encounter them.*

If the R-squared value is...	Then the correlation between the two variables is...	Reasonable conclusions include...
Between 0.70 and 1.00	Strong correlation	The model is a <u>very good</u> fit. Changes in the X-variable <u>explain most of the variation</u> in the Y-variable. The rest of the variation in the Y-variable is due to other variables and/or randomness.
Between 0.30 and 0.69	Moderate correlation	The model is a <u>good</u> fit. Changes in the X-variable <u>explain some of the variation</u> in the Y-variable. The rest of the variation in the Y-variable is due to other variables and/or randomness. There <u>may be</u> other explanatory variables which you have not yet analyzed, which could better explain the variation of the Y-variable.
Between 0.00 and 0.29	Weak correlation	The model is <u>not</u> a good fit. Changes in the X-variable <u>explain a little of the variation</u> in the Y-variable. There <u>are very likely</u> other explanatory variables which you have not yet analyzed, which could better explain the variation of the Y-variable.

Example: Suppose a local fitness center has conducted a study, tracking the activity of 20 of its female members, to measure the relationship between an individual's average minutes of exercise per week and her body fat percentage.

- It is unlikely that there is a clear cause-and-effect relationship, because many other lifestyle variables could influence the dependent variable (such as caloric intake). It makes sense to present this information in a **scatterplot**.

- The minutes of exercise should be plotted on the **X-axis**, because the convention is to plot **independent variables** on the **X-axis**. Intuitively, minutes of exercise should contribute toward changes in the body fat percentage. (The reverse is much less likely to be true).

- The body fat percentage should be plotted on the **Y-axis**, because the convention is to plot **dependent variables** on the **Y-axis**. Intuitively, body fat percentage is likely to be the outcome. (The reverse is much less likely to be true).

The data analyst has created two versions of the scatterplot using the same data set. One scatterplot features a linear trendline, whereas the other scatterplot features an exponential trendline.

Scatterplot 1A: Linear Trendline Equation

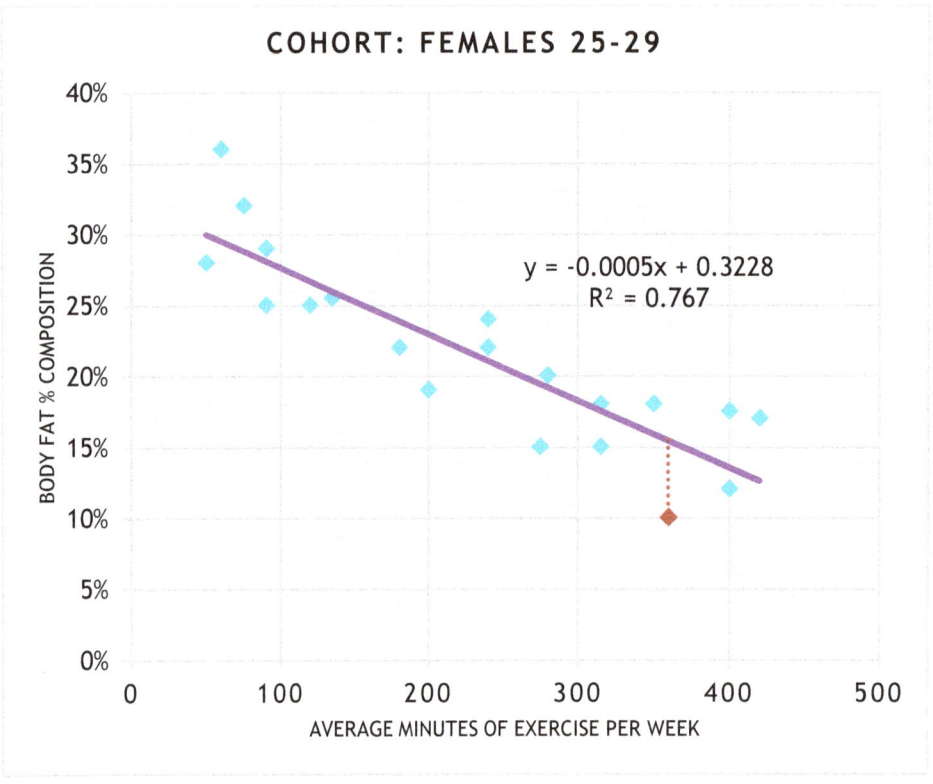

Scatterplot 1B: Exponential Trendline Equation

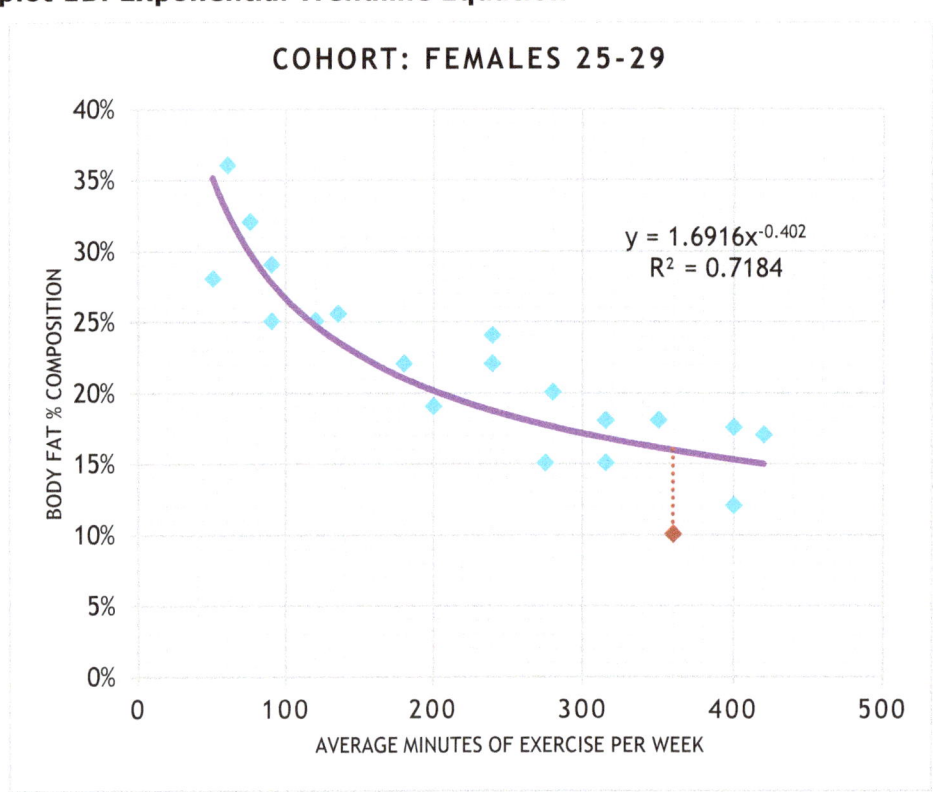

Interpreting Scatterplots, Trendline Equations, and R-squared values

- The **trendline** is indicated in purple.
- Each **marker** (here, the blue diamonds and single red diamond) indicates an individual value from the data set. There are 20 markers on the scatterplot, and thus 20 values in the data set.
    - The <u>vertical distance</u> between a single marker and the trendline indicates the difference between the <u>observed value</u> and the <u>predicted value</u>.
    - Look at the single red marker. It is <u>below</u> the **trendline**. Interpret this as "for this observation, body fat percentage was <u>less than expected</u> (<u>predicted</u>) by the trendline.
    - The length of the vertical, dotted red line indicates <u>by how much</u> the observed value fell below the expected value.
- The **trendline equation** and **R-squared value** are shown on each graph, above and to the right.
    - **Predicting values** using the **trendline equation**: Pay attention to whether you are given an X-value and asked to find the expected Y-value...or given a Y-value and asked what X-value would produce that result.
    - **Determining the better-fit trendline (model)**: Compare the R-squared values. R-squared values closer to 1.00 are a better fit.
        - In Scatterplot 1A, the R-squared value is 0.767
        - In Scatterplot 1B, the R-squared value is 0.7184
        - Thus, when using the **linear trendline** (model), changes in the average minutes of exercise per week **explain 76.7% of the variation** in an individual's body fat percentage.
        - Thus, when using the **exponential trendline** (model), changes in the average minutes of exercise per week **explain 71.84% of the variation** in an individual's body fat percentage.
        - The linear trendline is a **better-fit** than the exponential trendline, for this data set.
        - The rest of the variation in body fat percentage is due to either randomness or other explanatory variables not included in the model. *Those other variables could include the type of exercise performed, the intensity of exercise performed, dietary/nutrition choices, other health factors, genetics, and so on.*

PRACTICE SETS

SCATTERPLOT 1: STUDYING AND FINAL EXAM SCORES

A college economics professor, wishing to impress upon her students the impact of studying on their final exam scores, surveyed the students in her course the previous semester to ask, "How many minutes did you spend studying for the final exam?" and correlated this self-reported information with each person's actual score on the final exam.

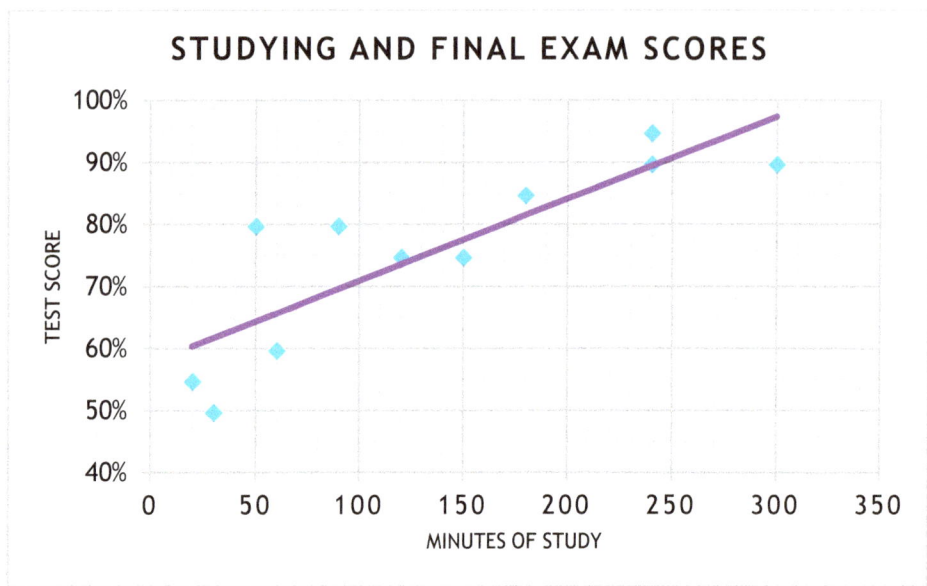

Question 1: *Refer to Scatterplot 1.* Which of the following statements are true, based upon the information shown? Choose all that apply.

   a) The student who most outperformed the expected exam score based on his or her minutes of studying spent 80 minutes studying.
   b) A total of five students performed worse on the final exam than expected, based on the number of minutes each spent studying.
   c) If a passing grade on the final exam is 70%, then three students failed the exam.

Question 2: *Refer to Scatterplot 1.* If a passing grade on the final exam is 80%, then the greatest number of minutes spent studying by a student who failed the exam is...

   a) 60
   b) 75
   c) 150

SCATTERPLOT 1: INTERPRETING THE DATA

A college economics professor, wishing to impress upon her students the impact of studying on their final exam scores, surveyed the students in her course the previous semester to ask, "How many minutes did you spend studying for the final exam?" and correlated this self-reported information with each person's actual score on the final exam.

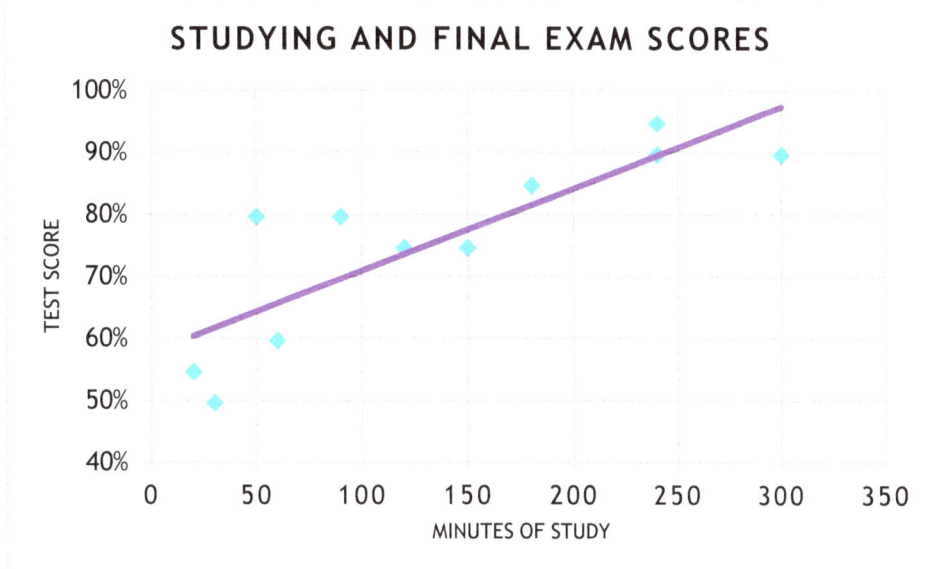

1. **Read the story** – the data relate to the minutes of studying and each student's final exam scores.

2. **Read the title** – Studying and final exam scores
   - When – no year is given, but the **story** mentions the data are from the "previous semester"
   - What – minutes of studying and final exam scores.
   - Who/Where – this professor's own students

3. **Read the labels** and **note the units of measurement for each variable**.
   - The X-axis label indicates this axis shows the minutes of study. It is scaled from 0 to 350 in major increments of 50. Each minor increment is 10, though these are not labeled.
   - The Y-axis label indicates this axis shows the test scores. It is scaled from 40% to 100% in major increments of 10%. Each minor increment is 5%, though these are not labeled.

4. **Read the legend** – not applicable

5. **Get a sense of the data** without dwelling on the details. Focus on the big picture.
   - In general, as the minutes of study increase, the final exam scores increase.

- There is a trendline, but the trendline equation is not included on the graph. *You should anticipate that this could be a question on a standardized test. Whenever there is a trendline but no equation, you may be asked which of several trendline equations best fits the data shown.*
- The data points on the left are further from the trendline, compared to the data points on the right.

6. **Read any footnotes and/or explanations** – not applicable

SCATTERPLOT 1: ANSWERS & EXPLANATIONS

Question 1: *Refer to Scatterplot 1.* Which of the following statements are true, based upon the information shown? Choose all that apply.

a) The student who most outperformed the expected exam score based on his or her minutes of studying spent 80 minutes studying.
b) A total of five students performed worse on the final exam than expected, based on the number of minutes each spent studying?
c) If a passing grade on the final exam is 70%, then three students failed the exam.

- Evaluate Statement A:
  - "The student who most outperformed" → In other words, lies farthest above the trendline
  - "the expected exam score based on his or her minutes of studying" → Recall that the expected exam score is the one predicted by the trendline
  - "spent 80 minutes studying" → The minutes are plotted on the X-axis.
  - Try putting this into your own words:
    - Which marker on the scatterplot has the greatest vertical distance between it and the trendline?
    - You can see that one of the markers is significantly above the trendline – the third one from the left.
    - What are the (X, Y) coordinates of that observation?
    - 50 minutes of studying and a score of 80%.
  - **Statement A is false**.
- Evaluate Statement B:
  - "A total of five students" → This is asking for the number of observations, so simply count the number of markers which meet the criteria (which come next)
  - "performed worse on the final exam than expected" → In other words, fell below the trendline
  - Try putting this into your own words:
    - How many markers on the scatterplot fall below the trendline?
    - 5 of them are clearly below the trendline. 2 appear to be on the trendline itself.
  - **Statement B is true**.
- Evaluate Statement C:
  - "If a passing grade on the final exam is a 70%" → This is not really a question; this phrase tells you that the cutoff for passing the exam is 70%.

- - - "then three students failed the exam" → Three students scored below a 70%.
    - Try putting this into your own words:
      - How many markers on the scatterplot fall below a Y-axis value of 70%?
      - 3 of them are below 70%.
    - **Statement C is true.**
- **You must choose statements B and C, but not A, to get the point for the question.**

Question 2: *Refer to Scatterplot 1. If a passing grade on the final exam is 80%, then the greatest number of minutes spent studying by a student who failed the exam is...*

   a) 60
   b) 75
   c) 150

- Recognizing the type of question: **Quantity comparison**.
- Solving the question: Use **Words to Math**.
    - "If a passing grade on the final exam is 80%" → This is not really a question; this phrase tells you that the cutoff for passing the exam is 80%. So, an 80% is a passing score and 79% or below is failing.
    - "Then the greatest number of minutes spent studying" → Look at the largest of the X-axis values, i.e., those furthest to the right
    - "by a student who failed the exam" → In other words, among students who scored less than an 80%
    - Try putting this into your own words:
        - Find all the students who failed. These are the data markers with a Y-value of less than 80%. (There are five of them).
        - Among only these five data markers, which one is furthest to the right? The one located at (150 minutes, 75%).
    - Choose C.
- **Behind each wrong answer choice is faulty logic**:
    - Choice A is incorrect. It is the result of assuming that the passing score in this question is the same as it was in Question 1. If you chose this answer, you may have worked too quickly and glossed over that detail in this question's wording.
    - Choice B is incorrect. This is the greatest <u>score</u> of the students who failed the exam. It is the result of misreading the two axes or working too quickly.

Off the Charts! Data Interpretation

SCATTERPLOT 2: URBAN COMMUTE TIMES

A city planner gathered data from a dozen cities to identify the relationship between the population and average one-way commute times. That information is presented in the chart below.

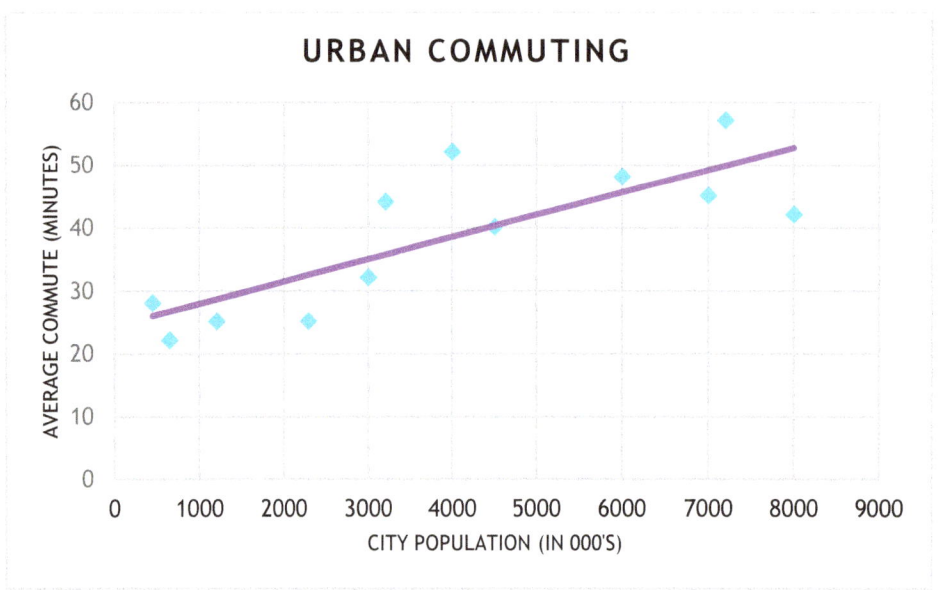

Question 1: *Refer to Scatterplot 2.* What is the least population among the cities with the three greatest average commute times?

a) 52
b) 4000
c) 7000
d) 4 million
e) 7 million

Question 2: *Refer to Scatterplot 2.* Based upon the information in the scatterplot, compare the following quantities:

Quantity A	Quantity B
The number of cities with an average round-trip commute time less than 40 minutes	5

a) Quantity A is greater
b) Quantity B is greater
c) Quantity A and Quantity B are equal
d) Cannot be determined

10-156   *Off the Charts! Data Interpretation*

Question 3: *Refer to Scatterplot 2.* Which of the following statements are true, based upon the information in the scatterplot? Choose all that apply.

   a) The predicted average one-way commute time for a city with a population of 6 million people is approximately 45 minutes.
   b) An average round-trip commute time of less than 1 hour is most likely to occur in cities with less than 1.5 million people.
   c) The cities in the sample with populations in excess of 5 million people all had average one-way commute times in excess of 40 minutes.

Off the Charts! Data Interpretation

SCATTERPLOT 2: INTERPRETING THE DATA

A city planner gathered data from a dozen cities to identify the relationship between the population and average one-way commute times. That information is presented in the chart below.

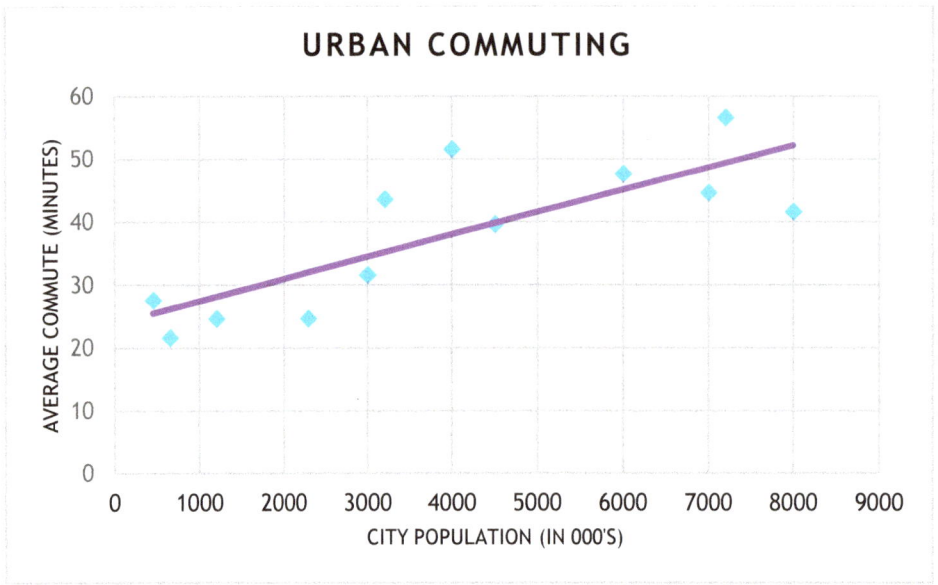

1. **Read the story** – the data relate to the population and one-way commute times of various cities.

2. **Read the title** – urban commuting
    - When – not provided in the **story** or **title**.
    - What – clarified in the **story**, these are one-way commute times as a function of population.
    - Who/Where – a dozen unnamed cities.

3. **Read the labels** and **note the units of measurement for each variable**.
    - The X-axis label indicates this axis shows the city population. It is scaled from 0 to 9000 in major increments of 1000. The text inside the parentheses after the axis label indicates the data are expressed in 000s, which means that three zeroes have been trimmed from each answer choice. Thus, an X-axis value of 4,000 really means 4,000,000 (4 million).
    - The Y-axis label indicates this axis shows the average commute, in minutes. It is scaled from 0 to 60 in major increments of 10 minutes.

4. **Read the legend** – not applicable

5. **Get a sense of the data** without dwelling on the details. Focus on the big picture.
    - In general, as the population increases, the average commute times increase.
    - There is a trendline, but the trendline equation is not included on the graph. *You should anticipate that this could be a question on a*

Off the Charts! Data Interpretation

standardized test. Whenever there is a trendline but no equation, you may be asked which of several trendline equations best fits the data shown.

- A few data points are further above/below the trendline than others, especially the cities with populations of 4 million and 8 million.

6. **Read any footnotes and/or explanations** – not applicable

SCATTERPLOT 2: ANSWERS & EXPLANATIONS

Question 1: *Refer to Scatterplot 2.* What is the least population among the cities with the three greatest average commute times?

a) 52
b) 4000
c) 7000
d) 4 million
e) 7 million

- Recognizing the type of question: **Statistical interpretation**. Find the **minimum** value within a subset of the data points.
- Solving the question: Use **Words to Math**.
    - "What is the least population" → Choose the smallest value for population
    - "among the cities with" → Analyze just a subset of the data points
    - "the three greatest average commute times" → This phrase describes the subset you should choose.
    - Paraphrasing this question:
        - Find the data points with the 3 greatest average commute times. *Commute time is plotted on the Y-axis, so look for the 3 points closest to the top of the graph and circle them.*
        - Among these 3 points, which has the smallest population? *Population is plotted on the X-axis, so determine which of the 3 points you circled is furthest to the left. Mark this point.*
        - What is the population for that point? *For the point you marked, drop straight down to the X-axis. This point has an X-coordinate of 4000. But, be careful! The X-axis label indicates that the population is "in 000s" so your "4000" is really "4000000."*
    - **Choose D**.
- **Behind each wrong answer choice is faulty logic**:
    - Choice A is incorrect, because it is the average commute time of the city which fits the criteria presented in the question. If you chose A, you may have done all of the initial steps correctly to identify the correct data point but read the value from the wrong axis. 52 is the Y-axis value.
    - Choice B is incorrect, because it fails to combine the information from the X-axis with the information from the X-axis label to scale the number appropriately. If you chose B, you may have neglected to read the X-axis label or failed to combine it with the numeric value from the X-axis.
    - Choice C is incorrect for two reasons. First, it results from a misinterpretation of the wording and selecting the three data points which have the greatest populations, rather than the greatest average commute times. Second, it fails to combine the information from the X-axis with the information from the X-axis label to scale the number appropriately.

- Choice E is incorrect, because it results from a misinterpretation of the wording and selecting the three data points which have the greatest populations, rather than the greatest average commute times.

Question 2: *Refer to Scatterplot 2.* Based upon the information in the histogram, compare the following quantities:

Quantity A	Quantity B
The number of cities with an average round-trip commute time less than 40 minutes	5

a) Quantity A is greater
b) Quantity B is greater
c) Quantity A and Quantity B are equal
d) Cannot be determined

- Recognizing the type of question: For questions which ask you to compare to quantities (typically found on the GRE®), you may need to translate one or two statements from words to math. Sometimes, the second quantity is simply a number (such as zero or 20) which requires no translation.
- Solving the question: Use **Words to Math** and **Words First, Math Second**.
- Evaluate Quantity A:
    - "The number of cities" → Count the data points which fit the criteria.
    - "with an average round-trip commute time" → Notice the wording change here! Based on the story above the scatterplot, the data show the one-way commute times. Round-trip commute times would be double the one-way commute times.
    - "less than 40 minutes" → Look for values which fall below this cutoff.
    - Determine your least-effort strategy:
        - You could double each of the values on the scatterplot, but this is time-consuming.
        - Instead, divide the value of 40 by 2, and count the number of cities with one-way commute times of less than 20 minutes.
    - None of the cities have average one-way commute times of less than 20 minutes, so none will have average round-trip commute times of less than 40 minutes.
- Compare your answer for Quantity A to the number in Quantity B.
- **Quantity B is greater**.
- **Choose B**.

The most common incorrect answer for this GRE-style quantity comparison question would be to choose C, that the two quantities are equal. If you chose C, you may have

10-162   *Off the Charts! Data Interpretation*

tried to save time by not reading the **story** above the scatterplot, and thus missed the key detail that the scatterplot shows one-way commute times and not round-trip commute times. *Never skip reading the story above any data visualizations. These stories contain valuable information.*

Question 3: *Refer to Scatterplot 2.* Which of the following statements are true, based upon the information in the scatterplot? Choose all that apply.

   a) The predicted average one-way commute time for a city with a population of 6 million people is approximately 45 minutes.
   b) An average round-trip commute time of less than 1 hour is most likely to occur in cities with less than 1.5 million people.
   c) The cities in the sample with populations in excess of 5 million people all had average one-way commute times in excess of 40 minutes.

- Recognizing the type of question: Each statement may require a different approach, so you'll need to use the same general process but different specific steps.
- Solving the question: Use **Words to Math** and **Words First, Math Second**.
- Evaluate Statement A:
    - "The predicted average one-way commute time" → The keyword <u>predicted</u> tells you to look at the **trendline**.
    - "for a city with a population of 6 million people" → Look at the X-axis. Find 6 million people. Because the X-axis is marked (in 000s), you will need to remove three zeros from 6 million to convert this to an X-axis value of 6000. Then, move up to the trendline, and to the left to find the Y-axis coordinate for this point on the trendline.
    - "is approximately 45 minutes" → Confirm whether the value you found in the previous step is equal to, or nearly equal to, 45 minutes.
    - Evaluate the truthfulness of the statement.
        - The Y-value of the trendline point corresponding to a population of 6 million people falls about halfway between the two major horizontal gridlines for 40 minutes and 50 minutes, so this is a good-enough answer.
        - **Statement A is true**.
- Evaluate Statement B:
    - "An average round-trip commute time" → Notice the wording change here! Based on the **story** above the scatterplot, the data show the <u>one-way</u> commute times. <u>Round-trip</u> commute times would be double the one-way commute times.
    - "of less than 1 hour" → Convert this to 60 minutes, because the Y-axis values are measured in minutes, not hours.
    - "is most likely to occur in cities with less than 1.5 million people" → The keyword <u>likely</u> tells you to interpret this as probability. You may not need to calculate an *exact* probability if the statement is either always true or always false.
    - Paraphrasing this question:
        - What is the population of cities which have one-way commute times of less than 30 minutes (round-trip commute times of less than 1 hour)?

- If these populations are all less than 1.5 million, then this statement is always true. If more than half of them are less than 1.5 million, then this statement is still true.
    - Evaluate the truthfulness of the statement.
        - There are 4 data points with one-way commute times of less than 30 minutes. 3 of these 4 data points have populations less than 1.5 million, which is more than half of the data points in this subset.
    - **Statement B is true**.
- Evaluate Statement C:
    - "The cities in the sample with populations in excess of 5 million people" → Look at the subset of data points with populations greater than 5 million. Do not include those with populations of exactly 5 million. Look at the data points to the right of the vertical gridline for 5 million (5000). Circle the data points in this subset.
    - "all had average one-way commute times" → Be careful when you see words such as <u>all</u> or <u>none</u>.
    - "in excess of 40 minutes" → Greater than 40 minutes. Do <u>not</u> include those with commute times of exactly 40 minutes. Confirm whether all of the data points are above the horizontal gridline for 40 minutes.
    - All of the data points in the subset are above the horizontal gridline for 40 minutes.
    - **Statement C is true**.
- **You must choose all three statements A, B, and C, to get the point for the question**.

CHAPTER 11 PAIRED & MIXED DATA VISUALIZATIONS

CONCEPT REVIEW

When you need to provide either information about more variables, different perspectives on the same data, or greater detail about the data, you may find it challenging to fit all of this information into a single data visualization that is still comprehensible to the people consuming the information.

There's a dynamic tension among:

- The amount of space available on the page or screen
- The desire to convey precise information and more detail
- The audience confusion that can result from too much "visual clutter"
    - Which is related to the relative sophistication (or lack of sophistication) of your intended audience. *You cannot control your audience's sophistication with data, but you can accommodate varying levels of fluency when you design your charts and graphs so that you influence how your information is perceived.*
    - Which is also related to such visual elements as white space, color, font size, callouts, and other factors which impact readability. *You can influence your audience's interpretation of data visualizations through thoughtful use of these factors. Skillful use of these visual elements meaningfully impacts how readily your audience comprehends the information and insights found in all types of data visualizations. That said, the ones you'll see on standardized tests are not designed to make the insights obvious, but to* **test your ability to apply your reasoning skills and figure out the insights***.*

Paired data visualizations and **mixed data visualizations** are ways of accomplishing the same goal – providing more information about a given phenomenon (business performance, public health trends, urbanization patterns, et al) in a way which facilitates comprehension of the relationships among multiple variables.

- Throughout this book, the term **paired data visualizations** refers to the use of two (or more) distinct but related data visualizations. Depending upon the orientation of the page or screen, **paired data visualizations** may be presented side-by-side or stacked (above-and-below).
- The term **mixed data visualizations** refers to the overlay of two different types of graphs, charts, or other visualizations into a single data visualization.

Both types present more information than a single data visualization could, and thus result in greater complexity.

	Paired Data Visualizations	**Mixed Data Visualizations**
Benefits	<ul><li>Can help keep the two concepts distinct</li><li>Can avoid some of the visual "clutter" which makes viewers feel overwhelmed</li><li>May enable the <u>creator</u> to enlarge font sizes, thereby improving readability</li></ul>	<ul><li>Take up less space on the page or screen</li><li>The <u>creator</u> may be able to make it clearer how the mixed visualizations are related (fit together)</li></ul>
Downsides	<ul><li>Take up more space on the page or screen</li><li>The <u>viewer</u> must figure out how the two visualizations are related (fit together)</li></ul>	<ul><li>Can create "visual clutter" which makes viewers feel overwhelmed</li><li>May force the creator to shrink font sizes, thereby impeding readability, or suppress some of the data labels, to aid readability</li><li>Harder for viewers to interpret, as mixed data visualizations have <u>more layers of information</u> and may require a 2nd Y-axis</li></ul>
Example and Purpose	EXAMPLE<ul><li>A pair of pie charts, both showing the breakdown of a college's enrollment by the students' major for two different schools.</li></ul>PURPOSE<ul><li>To <u>facilitate comparison</u> of the two schools, while also allowing analysis of each school separately.</li><li>The paired pie charts would make it easy to see that one school has a <u>greater proportion</u> of a certain major than the other school.</li></ul>	EXAMPLE<ul><li>A column chart showing revenue over time, overlaid with a line chart showing profit percentage. This chart would have two Y-axes, one on the left (revenue) and one on the right (profit %).</li></ul>PURPOSE<ul><li>To <u>show (both separately and combined) the trends</u> of revenue and profit percentage, which do not always move in tandem.</li><li>To <u>highlight when trends converge or diverge</u> for the different variables.</li></ul>

Off the Charts! Data Interpretation

	Paired Data Visualizations	**Mixed Data Visualizations**
Example and Purpose	EXAMPLE • A pie chart, showing a store's revenue by product category, paired with a bar chart which shows the sales of individual products within a single category. PURPOSE • To show the overall proportion of the major categories, while providing extra detail about a single category. • Pie charts are limited in their usefulness when categories exceed the ideal range of 2-to-7 categories. Often, that last category might be labeled "all other." • The drill-down in the second chart enables a more detailed analysis. For example, the second chart provides a breakdown of the "all other" category. • Putting that level of detail into a single pie chart would have made each slice of the pie chart too small to be usefully understood (e.g., 5 product categories but 37 different products).	EXAMPLE • A stacked bar chart showing the store's revenue by product category over time. • The overall height of each bar would show total sales in that time period. • The height of each section of the bars would show the sales of a specific product category in that time period. PURPOSE • To show the trend of how total revenue changed over time using the total height of each bar. • To show the trend of how revenue from each category changed over time using the height of same-colored sections over time. • To show the proportion of revenue by category within any single time period. • In summary, stacked bar charts are like combining multiple pie charts in a single graphic, while allowing you to see trends.

Whenever you encounter **paired data visualizations**, you should immediately look over the core elements of each data visualization (title, axis titles, headers/labels, etc.) and ascertain:

- What elements are the same?
- What elements are different?

Then, seek to articulate how the **paired data visualizations** are related. Some common relationships between paired data visualizations include:

Relationship Pattern	Examples
The same variables measured at different locations	• Incidence rates of various food allergies in Country A vs. Country B. • Production costs at Factory A vs. Factory B. • Unemployment in City A vs. City B.
The same variables measured for different categories	• Jobs created in High Tech vs. Low Tech industries. • Worldwide demand for a certain oil, by country, for use as either an ingredient or an end product.
Category-and-subcategory (group-and-segment) detail	• Number of new jobs created in City A, categorized by industry, with more detail about the number of jobs in various salary bands for Industry X. • Number of patients seen in the emergency room at a certain hospital, by type of doctor seen, with more detail about the frequency of various injuries seen by the orthopedic team. • Amount of added sugar found in foods marketed to children, with more detail about the amount of added sugar in specific kinds of cereal.
The same variables measured at different time periods	• Number of professors hired into each of a University's 12 colleges, in 2000 versus 2010. • Proportion of students enrolling in each of several majors, in 1980 versus 2010. *Note: if you are asked to calculate the average growth per year, be sure to correctly calculate based on elapsed time. You may need to look at the month and day. For example, three full years elapse from January 1, 2014 to December 31, 2016.

Relationship Pattern	Examples
The <u>same population</u>, but <u>different variables</u>	- A population might be grouped into 4 age segments, with one chart showing the average 5K race time and another chart showing the average amount of sleep.
Two <u>parts of one process</u>	- The amount of time that beginners need to advance to intermediate level, and the amount of time that intermediate level individuals need to advance to expert level. - The amount of time needed to <u>setup</u> a factory for production of Good X, and the amount of time needed to <u>complete the production</u> run of Good X. - The reaction time of a driver and braking times of a race car, as a function of a car's velocity. The reaction time and braking time would be combined to find the total stopping time.

PRACTICE SETS

PAIRED CHARTS 1: HOUSEHOLD SPENDING DATA

The government of Country X gathered data about median household spending in 2014 and 2015. Use the following charts to answer the questions below.

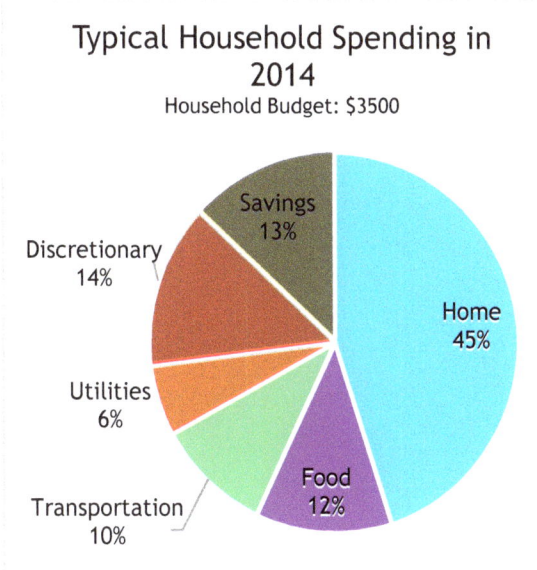
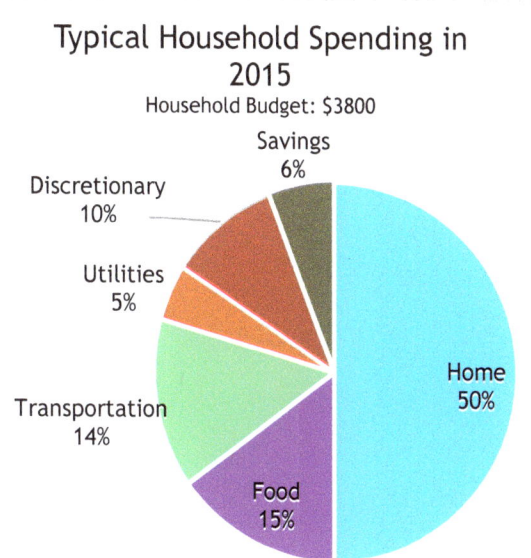

Question 1: *Refer to Paired Charts 1.* Did the typical household spend more on Utilities in 2014 or in 2015?

   a) 2014
   b) 2015
   c) The same in 2014 & 2015
   d) Cannot be determined from the information given

Question 2: *Refer to Paired Charts 1.* The amount the typical household spent on their home increased by approximately what percent from 2014 to 2015?

   a) 5%
   b) 11%
   c) 17%
   d) 21%

Off the Charts! Data Interpretation

Question 3: *Refer to Paired Charts 1.* The amount the typical household spent on food and transportation was approximately what percent less in 2014 than the amount spent on food and transportation in 2015?

a) 7%
b) 24%
c) 30%
d) 43%

PAIRED CHARTS 1: INTERPRETING THE DATA

The government of Country X gathered data about median household spending in 2014 and 2015. Use the following charts to answer the questions below.

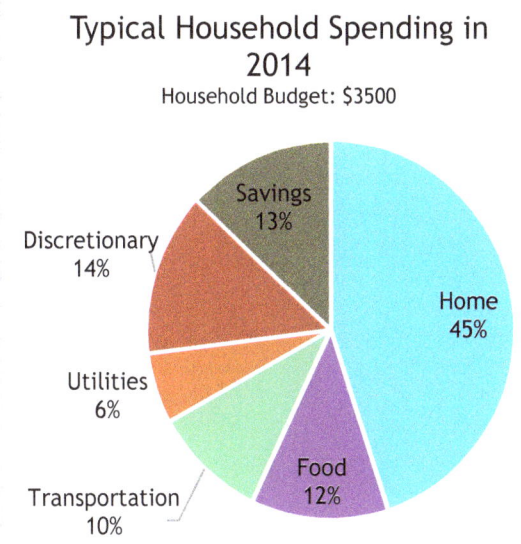
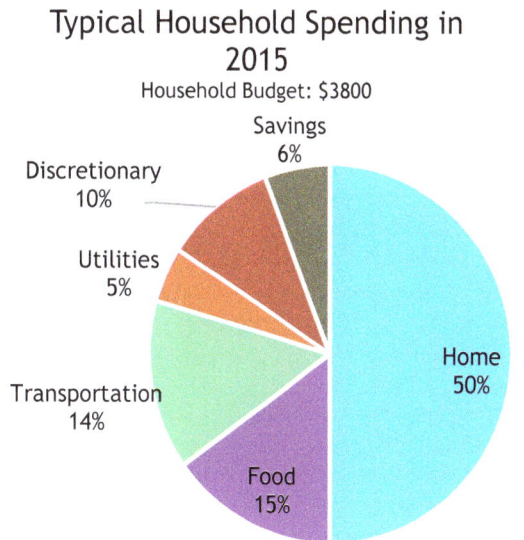

1. **Read the story** – from this, you learn that the data are about spending data for one country with six different spending categories, in two different years.

2. **Read the title** – Typical Household Spending

    - When – the pie chart on the left is for 2014. The pie chart on the right is for 2015.

    - What – Typical Household Spending. The **story** tells you this is the **median** data (so don't mistake that for <u>average</u>)

    - Who/Where – not shown in the title, but the **story** tells you this is for Country X

    - Subtitle – notice that in the small print, you learn that the size of the pie changed. The household budget in 2014 was $3500, and in 2015 it was $3800.

3. **Read the labels** and **note the units of measurement for each variable**.

    - Both pie charts show the same categories with relative numbers in percentages.

    - *Whenever you see paired pie charts displaying percentages and the size of the pie is different, you should anticipate one of the questions will require you to combine this information.*

4. **Read the legend** – not applicable

5. **Get a sense of the data** without dwelling on the details. Focus on the big picture.

Off the Charts! Data Interpretation

- The percentages spent on Home, Food, and Transportation each increased.
- The percentage allocated to Savings shrank significantly.
- The percentage spent on Utilities decreased, but the size of the pie (the total budget) grew. *Whenever you have a **mixed effect** like this, you should anticipate one of the questions will ask you about the percent change for this category.*

6. **Read any footnotes and/or explanations** – not applicable
7. **Identify how the two data visualizations are related**.
    - The second pie chart offers you a comparison to the first pie chart, to see how the spending allocation changed from one year to the next.
    - The variables and categories are otherwise the same.
    - **The "size of the pie" differs**. The total budget is written in small print below the title of each chart. As mentioned above, anticipate a question about this information.

PAIRED CHARTS 1: ANSWERS & EXPLANATIONS

Question 1: *Refer to Paired Charts 1.* Did the typical household spend more on Utilities in 2014 or in 2015?

   a) 2014
   b) 2015
   c) The same in 2014 & 2015
   d) Cannot be determined from the information given

- Recognizing the type of question: **Absolute comparisons about two values**. It's asking you to determine which value is larger and when that happened.
- Be careful – the percentage decreased from 2014 to 2015, but the size of the pie increased. Because there is a **mixed effect**, take the percentage from each year times the total amount in each year.
    - *Amount spent on Utilities in* $2014 = 6\% * \$3500 = $**$210**
    - *Amount spent on Utilities in* $2015 = 5\% * \$3800 = $**$190**
- The typical household spent more in 2014.
- Choose A.

Question 2: *Refer to Paired Charts 1.* The amount the typical household spent on their home increased by approximately what percent from 2014 to 2015?

a) 5%
b) 11%
c) 17%
d) 21%

- Recognizing the type of question: **Part-to-part comparison**. Specifically, it's asking for the **percentage change** info about the home category, from one year to the next.
- Solving the question: Use **Words to Math** and **Words First**, **Math Second**.
    - "The amount the typical household spent on their home" → examine the pie chart and carefully identify the home amount. It's not there. Instead, you have a home percentage. You'll need to calculate the rent amount by taking the home percentage times the total amount.
    - "increased by approximately what percent" → use the percent change equation.
    - $Percent\ Change = \dfrac{New - Original}{Original}$
    - "from 2014" → 2014 is our "original"
    - "to 2015" → 2015 is our "new"
    - $Percent\ Change = \dfrac{2015\ Home\ Amount - 2014\ Home\ Amount}{2014\ Home\ Amount}$
    - $Percent\ Change = \dfrac{(2015\ Home\ \% * 2015\ Total) - (2014\ Home\ \% * 2014\ Total)}{(2014\ Home\ \% * 2014\ Total)}$
    - $Pct.\ Change = \dfrac{(50\% * \$3800) - (45\% * \$3500)}{(45\% * 3500)} = \dfrac{1900 - 1575}{1575} = \dfrac{325}{1575} = \mathbf{20.6\%}$
    - Choose D, 21%.
- **Behind each wrong answer choice is faulty logic**:
    - Choice A is incorrect because it fails to account for the "percent change of a percent" relationship.
    - Choice B is incorrect. Though the basic percent change calculation was executed correctly, this answer fails to account for how the size of the pie changed.
    - Choice C is incorrect. This answer correctly accounts for how the size of the pie changed, but results from incorrectly executing the percent change formula by putting the wrong value in the denominator.

Question 3: *Refer to Paired Charts 1. The amount the typical household spent on food and transportation was approximately what percent less in 2014 than the amount spent on food and transportation in 2015?*

   a) 7%
   b) 24%
   c) 30%
   d) 43%

- Recognizing the type of question: **Subtotal-to-subtotal comparison**. It's asking for the **percentage change** of the combined spending on Food + Transportation.
- Solving the question: Use **Words to Math** and **Words First, Math Second**.
  - "The amount the typical household spent on food and transportation" → examine the pie chart and carefully identify the food and transportation amount. It's not there. Instead, you have a percentage for food and one for home. You'll need to add these percentages, then multiply the combined percentages by the total amount to find the combined amount
  - "was what percent less" → use the percent change equation.
  - $Percent\ Change = \frac{Focus - Basis\ of\ Comparison}{Basis\ of\ Comparison}$
  - "in 2014" → 2014 is the "focus"
  - "than the amount spent on food and transportation"
  - "in 2015" → 2015 is the "basis of comparison"
  - $Percent\ Change = \frac{2014\ Food\ \&\ Trans\ Amt - 2015\ Food\ \&\ Trans\ Amt}{2015\ Food\ \&\ Trans\ Amt}$
  - $Pct\ Change = \frac{[(2014\ Food\% + Trans\%)*2014\ Tot.] - [(2015\ Food\% + Trans\%)*2015\ Tot.]}{[(2015\ Food\% + Trans\%)*2015\ Tot.]}$
  - $Percent\ Change = \frac{[(12\% + 10\%)*\$3500] - [(15\% + 14\%)*\$3800]}{[(15\% + 14\%)*3800]}$
  - $Percent\ Change = \frac{(22\%*\$3500) - (29\%*3800)}{(29\%*3800)} = \frac{770 - 1102}{1102} = \frac{-332}{1102} = \mathbf{30\%}$
  - Choose C, 30%.
- **Behind each wrong answer choice is faulty logic**:
  - Choice A is incorrect because it fails to account for the "percent change of a percent" relationship.
  - Choice B is incorrect because it makes two mistakes. It does not modify the percent change formula to show that the size of the pie changed. The change was divided by the wrong denominator. When you compare 2014 to 2015, then 2015 is the basis of comparison and thus the correct denominator.
  - Choice D is incorrect. The basic percent change calculation correctly reflected that the size of the pie changed, but the change was divided by the wrong denominator. When you compare 2014 to 2015, then 2015 is the basis of comparison and thus the denominator.

PAIRED CHARTS 2: BOUTIQUE SALES

The owner of a retail store which sells clothing, accessories, and shoes has provided you the following information about the breakdown of the store's revenues for the first half of last year. The owner then asks you several questions.

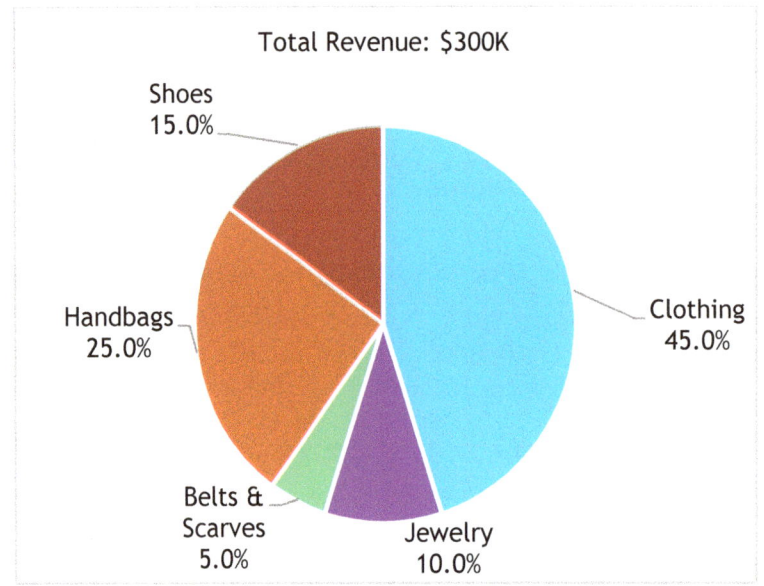

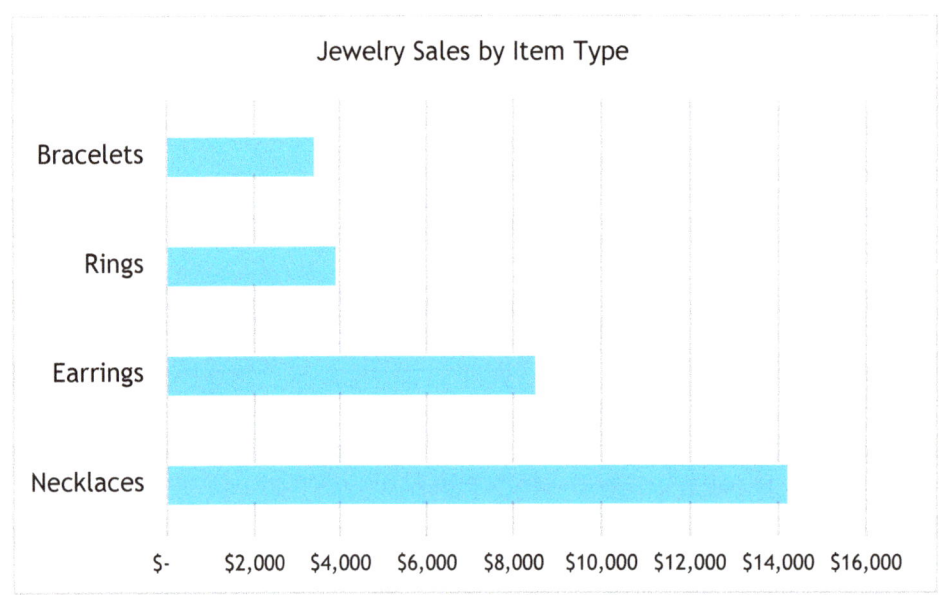

Question 1: *Refer to Paired Charts 2.* Sales of earrings constituted approximately what percent of total sales for the first half of last year?

   a) 2.8%
   b) 10.0%
   c) 28.3%

Question 2: *Refer to Paired Charts 2.* If the amount of handbags revenue tripled in the second half of the year, and sales of all other items remained the same, then handbags would account for approximately what percent of total revenue for the full year?

   a) 25%
   b) 40%
   c) 50%
   d) 75%

Off the Charts! Data Interpretation

PAIRED CHARTS 2: INTERPRETING THE DATA

The owner of a retail store which sells clothing, accessories, and shoes has provided you the following information about the breakdown of the store's revenues for the first half of last year. The owner then asks you several questions.

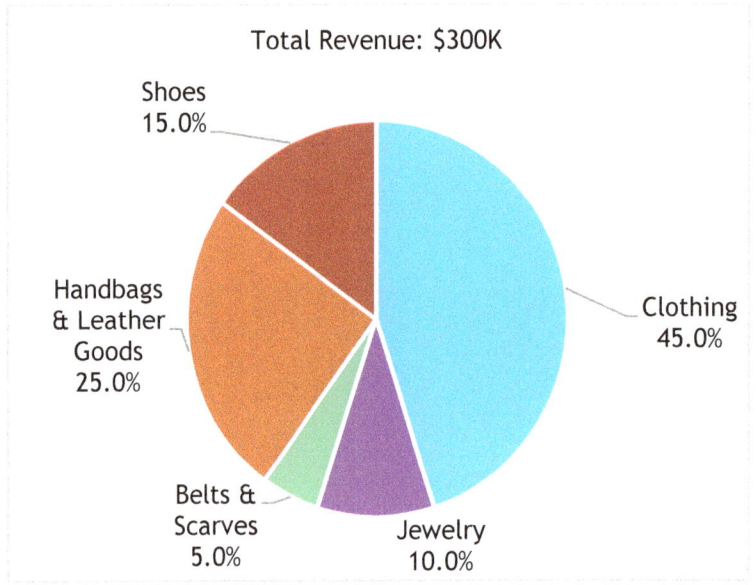

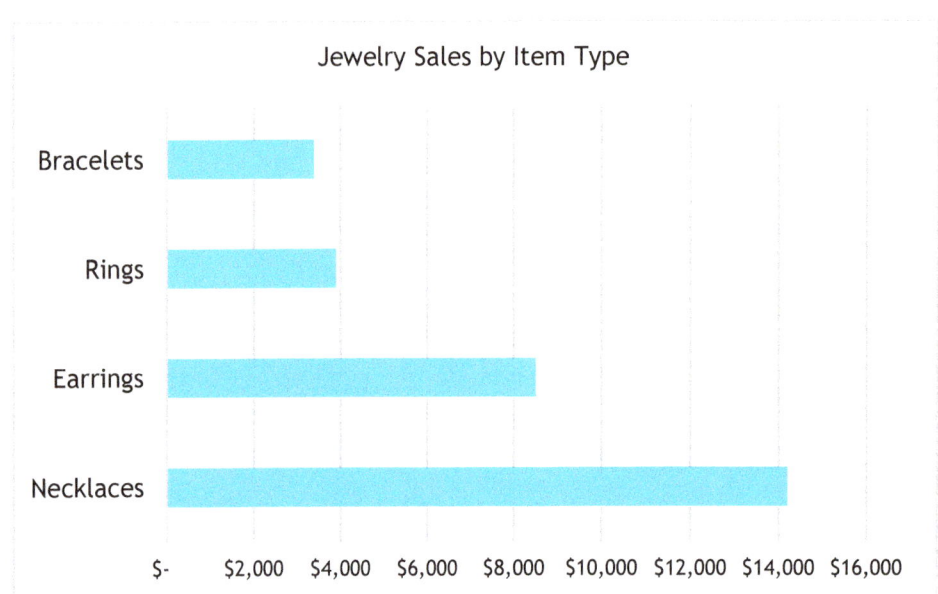

1. **Read the story** – from this, you learn that the data are about a store's revenue.
2. **Read the title** – Total Revenue: $300K and Jewelry Sales by Item Type
   - *You can hypothesize, but not yet conclude, that the second chart provides more detail about a **segment** of the first chart.*
   - When – according to the **story**, the revenue data are from the first half of last year.

11-180   *Off the Charts! Data Interpretation*

- *You might anticipate that there could be a question about the store's full year revenues.*
- What –
    i. Revenue by Product Category. The **story** plus the **data labels** on the pie chart allow you to infer these are the store's product categories.
    ii. Jewelry sales by item type.
    iii. If you quickly add up the values from the 2nd chart, then compare this sum with jewelry % of total revenue from the 1st chart, you find these are the same value, $30,000.
    iv. *Now you can confirm that the second chart is providing full detail about a **segment** of the first chart.*
- Who/Where – not shown in the title, but the **story** tells you this is for a store

3. **Read the labels** and **note the units of measurement for each variable**.
    - The pie chart shows 5 product categories and each one's percentage of the total.
    - The bar chart shows 4 kinds of jewelry, which is one of the product categories from the pie chart.

4. **Read the legend** – not applicable

5. **Get a sense of the data** without dwelling on the details. Focus on the big picture.
    - Clothing is the store's highest-revenue product category.
    - There are four types of jewelry within the jewelry product category.
    - Revenue from the sale of bracelets & rings are close in value.

6. **Read any footnotes and/or explanations** – not applicable

7. **Identify how the two data visualizations are related**. Already done, above.

PAIRED CHARTS 2: ANSWERS & EXPLANATIONS

Question 1: Refer to Paired Charts 2. Sales of earrings constituted approximately what percent of total sales for the first half of last year?

a) 2.8%
b) 10.0%
c) 28.3%

- Recognizing the type of question: **Part-to-total comparison**. Specifically, it's asking for the **percentage of** the total.
- Solving the question: Use **Words to Math** and **Words First, Math Second**.
    - "Sales of earrings constituted approximately" → Earrings are a type of jewelry, so this information is found in the bar chart. It appears to be about one-quarter of the way between the major gridlines of $8,000 and $10,000, so estimate this value as $8,500.
    - "what percent of" → This is your variable to solve for
    - "total sales for the first half of the year" → Look at the title of the first chart. The total sales amount is $300K, or $300,000.
    - Set up the equation and solve.
    - $Sales\ of\ Earrings = P\ percent * Total\ Sales\ 1st\ Half$
    - $\$8,500 = P\ percent * \$300,000$
    - $\frac{\$8,500}{\$300,000} = P\ percent = 0.028333\ or\ 2.83\%$
- Choose A.
- **Behind each wrong answer choice is faulty logic**:
    - Choice B is the percent of total sales for all jewelry. If you chose this one, you missed a step or misread the question.
    - Choice C is the result of mistakes related to place value (misplacing the decimal).

Question 2: *Refer to Paired Charts 2. If the amount of handbags revenue tripled in the second half of the year, and sales of all other items remained the same, then handbags would account for approximately what percent of total revenue for the full year?*

   a) 25%
   b) 40%
   c) 50%
   d) 75%

- Recognizing the type of question: **Predictions about new data points, based on trends**. For this one, you will need to **extrapolate** from the data in the chart about this store to a new value about the whole school district.
- Solving the question: Use **Words to Math** and **Words First, Math Second**.
    - "If the amount of handbags revenue tripled in the second half of the year" → The signal word "if" indicates you need to extrapolate.
        - Find the amount of handbags revenue in the 1st half = 25% * 300,000 total revenue = $75,000
        - Triple that = $225,000 in the 2nd half of the year
    - "and sales of all other items remained the same"
        - Find the sales of other items: Take the total $300,000 and subtract the handbags sales of $75,000 = $225,000 in the 1st half of the year. Sales of other items in the 2nd half are also $225,000.
        - Combined then, total sales in the 2nd half of the year = $450,000
    - "then handbags would account for approximately what percent of" → Set up a percentage calculation.
    - "total revenue for the full year" → This is the denominator. $300,000 in the first half of the year, plus $450,000 in the second half of the year.
    - $Pct\ of\ Full\ Year = \frac{Full\ Year\ Handbags}{Full\ Year\ Total\ Revenue}$
    - $Pct\ of\ Full\ Year = \frac{1st\ Half\ Handbags + 2nd\ Half\ Handbags}{1st\ Half\ Total + 2nd\ Half\ Total}$
    - $Pct\ of\ Full\ Year = \frac{\$75,000 + \$225,000}{\$300,000 + \$450,000} = \frac{\$300,000}{\$750,000} = \mathbf{40\%}$
- Choose B.
- **Behind each wrong answer choice is faulty logic**:
    - Choice A is incorrect. This is a random extra answer choice.
    - Choice C is incorrect. This answer choice expresses the percentage for the second half of the year, not the full year. If you chose this answer, you may have misread the question or not realized extra steps were needed.
    - Choice D is incorrect. You cannot simply triple the percentage, because changing the dollar amount of sales of handbags would also change the total, thereby changing all the percentages. If you chose this answer, you were likely rushing through the question.

MIXED CHARTS 1: STUDENT READING PROFICIENCY

An education policy specialist gathered information about the four school districts in his city regarding the rates of student reading proficiency among 6th grade students. For the purpose of this study, proficiency means that the student is performing at or above grade-level on the end of year reading comprehension exams.

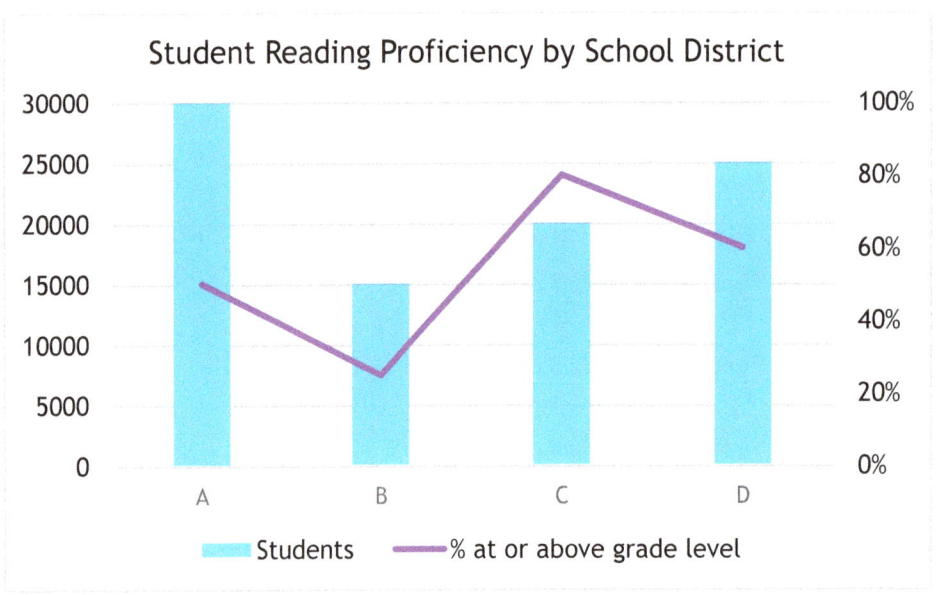

Question 1: Refer to Mixed Charts 1. Based upon the information in the graph, compare the following quantities:

Quantity A	Quantity B
The number of students who are reading proficiently in District A	The number of students who are reading proficiently in District D

a) Quantity A is greater
b) Quantity B is greater
c) Quantity A and Quantity B are equal
d) Cannot be determined

Question 2: Refer to Mixed Charts 1. The percentage of students who are not proficient in reading is lowest in which of the following districts?

a) District A
b) District B
c) District C
d) District D

Question 3: *Refer to Mixed Charts 1.* Which of the following is the best approximation of the citywide rate of reading proficiency?

- a) 40%
- b) 55%
- c) 66%
- d) 70%

MIXED CHARTS 1: INTERPRETING THE DATA

An education policy specialist gathered information about the four school districts in his city regarding the rates of student reading proficiency among 6th grade students. For the purpose of this study, proficiency means that the student is performing at or above grade-level on the end of year reading comprehension exams.

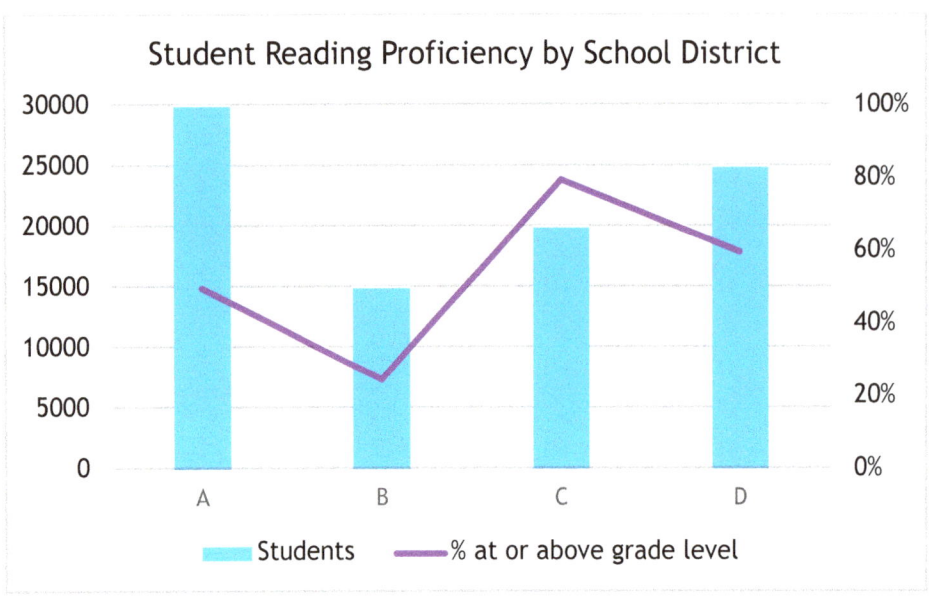

1. **Read the story** – from this, you learn that the data are about four school districts and their rates of student reading proficiency among 6th graders.

2. **Read the title** – Student Reading Proficiency by School District
    - When – not provided in the **story** or **title**
    - What –
        i. Number of students
        ii. % at or above grade level (from the story, you can infer this is synonymous with "at or above grade level")
    - Who/Where – four unnamed school districts

3. **Read the labels** and **note the units of measurement for each variable**.
    - The X-axis lists the four school districts in alphabetical order
    - The Y-axis on the left is a number, whereas the Y-axis on the right is a percentage.
        i. You can infer that the Y-axis on the left is the number of students. When you are reading the values from each column, go to the left.
        ii. You can infer that the Y-axis on the right is the percentage that are at or above grade level. When you are reading the values from the line, go to the right.
    - There are six gridlines.

- There are six non-zero values marked on the Y-axis on the left.
- There are five non-zero values marked on the Y-axis on the right.
- You can infer then that the gridlines are associated with the number of students on the left, not with the percentages on the right.
- Therefore, you may need to be extra careful when estimating the percentages which correspond to data points on the line graph.

4. **Read the legend** – the blue columns represent the number of students, and the purple line represents the percentage of these students who are reading at or above grade level.

5. **Get a sense of the data** without dwelling on the details. Focus on the big picture.
    - District A has the most students, whereas District B has the fewest students.
    - District B has both the smallest number of students and the smallest percentage reading proficiently, so District B definitely has the smallest number of students reading proficiently.
    - District C has the highest percentage reading proficiently.

6. **Read any footnotes and/or explanations** – not applicable

7. **Identify how the two data visualizations are related**. Already done, above.

Off the Charts! Data Interpretation

MIXED CHARTS 1: ANSWERS & EXPLANATIONS

Question 1: *Refer to Mixed Charts 1.* Based upon the information in the graph, compare the following quantities:

Quantity A	Quantity B
The number of students who are reading proficiently in District A	The number of students who are reading proficiently in District D

a) Quantity A is greater
b) Quantity B is greater
c) Quantity A and Quantity B are equal
d) Cannot be determined

- Recognizing the type of question: **Quantity Comparison**. Compare two values.
- Solving the question: Use **Words to Math** and **Words First, Math Second**.
- Evaluate Quantity A:
    - "The number of students who are reading proficiently" → This information is not provided directly in the mixed chart, which shows the number of students and the percentage reading at or above grade level.
        - You will need to combine the two types of data in the mixed chart to calculate the number of students who are reading proficiently.
        - $\# \, Reading \, Proficiently \, = \, \# \, of \, Students \, * \, \% \, Proficient$
    - "in district A" → Find District A on the X-axis and retrieve the necessary values from the column and line associated with District A.
        - $\# \, Reading \, Proficiently \, = \, 30{,}000 * \frac{50}{100} = \mathbf{15{,}000}$
- Evaluate Quantity B:
    - "The number of students who are reading proficiently" → This information is not provided directly in the mixed chart, which shows the number of students and the percentage reading at or above grade level.
        - You will need to combine the two types of data in the mixed chart to calculate the number of students who are reading proficiently.
        - $\# \, Reading \, Proficiently \, = \, \# \, of \, Students \, * \, \% \, Proficient$
    - "in district D" → Find District A on the X-axis and retrieve the necessary values from the column and line associated with District A.
        - $\# \, Reading \, Proficiently \, = \, 25{,}000 * \frac{60}{100} = \mathbf{15{,}000}$
- Compare Quantity A and Quantity B. **The two quantities are equal**.
- Choose C.

Question 2: *Refer to Mixed Charts 1.* The percentage of students who are not proficient in reading is lowest in which of the following districts?

   a) District A
   b) District B
   c) District C
   d) District D

- Recognizing the type of question: **Relative comparisons** and **statistical interpretations**. You want to find the **minimum** percentage. Here, you are asked to find the percentage who are <u>not</u> proficient. The line graph shows the percentage who <u>are</u> proficient.

- Because proficient and not proficient are binary conditions (meaning, there is no possible 3rd scenario), you simply retrieve the percentage who are proficient, and subtract that from 100%.

- You may find it helpful to create a simple table to help you organize the data:

District	100%	-	% Proficient	=	% Not Proficient
A	100%		50%	=	50%
B	100%		25%	=	75%
C	100%		80%	=	20%
D	100%		60%	=	40%

- Alternatively, you may realize that the district with the <u>lowest</u> percentage of students who <u>are not</u> proficient would be the <u>same</u> as the one with the <u>highest</u> percentage of students who <u>are</u> proficient. Look for the highest point on the line graph. That's District C.

- Choose C.

- **Behind each wrong answer choice is faulty logic**:

    o *Choice A is incorrect. This is a random extra answer choice.*

    o *Choice B is incorrect. This district has the lowest percentage of students who <u>are</u> proficient. If you chose this answer, you may have missed the critical word <u>not</u> when you read the question.*

    o *Choice D is incorrect. This is a random extra answer choice.*

Off the Charts! Data Interpretation

Question 3: *Refer to Mixed Charts 1.* Which of the following is the best approximation of the citywide rate of reading proficiency?

    a) 40%
    b) 55%
    c) 66%
    d) 70%

- Recognizing the type of question: This one is a big tougher to determine, because it requires you to use logic to interpret the meaning of "citywide rate" of reading proficiency. Go back to the **story** above the mixed chart, and you will see that there are four districts in the city. Thus, you can infer that the "citywide rate" should be the average rate of reading proficiency across the four districts. Because there are different numbers of students in each district, you must calculate a **weighted average** or **expected value**.
- You can either set up an equation or create a table to organize the data and calculations.
- You could use the weighted average equation to find the citywide rate of reading proficiency:
- $Weighted\ Average = \dfrac{(\#in\ A * \%\ A) + (\#in\ B * \%\ B) + (\#\ in\ C * \%\ C) + (\#\ in\ D * \%\ D)}{\#\ in\ A + \#\ in\ B + \#\ in\ C + \#\ in\ D}$
- $Weighted\ Average = \dfrac{(30000 * 0.50) + (15000 * 0.25) + (20000 * 0.80) + (25000 * 0.60)}{30000 + 15000 + 20000 + 25000}$
- $Weighted\ Average = \dfrac{(15000) + (3750) + (16000) + (15000)}{90000} = \dfrac{49750}{90000} = \mathbf{55\%}$
- You can also create a simple table to help you organize the data and ensure you are calculating the answer correctly:

District	# of Students	*	% Proficient	=	# Proficient
A	30,000	*	$\dfrac{50}{100}$	=	15,000
B	15,000	*	$\dfrac{25}{100}$	=	3,750
C	20,000	*	$\dfrac{80}{100}$	=	16,000
D	25,000	*	$\dfrac{60}{100}$	=	15,000
Total	90,000	*	$\dfrac{Total\ \#\ Proficient}{Total\ \#\ Students}$ $\dfrac{49{,}750}{90{,}000} = 55\%$	=	49,750

- Choose B.

CHAPTER 12 REFRESHER CONTENT & STRATEGIES

This section contains general strategies for working with word problems, plus selected content for the most common types of word problems about data visualizations.

> For more in-depth preparation focused on Word Problems, including practice questions with follow-along explanations, **check out the author's other book**, *All Your Word Problems Solved: Crushing Standardized Test Math*.

TRANSLATING WORDS TO MATH

When you're handling word problems, either for classes or for one of the standardized college and graduate admissions tests (PSAT/NMSQT, SAT, ACT, GRE, GMAT, etc.), there are a few techniques you can use which make it easier to translate words to math:

1. **Read the first sentence or so of the problem**, <u>just enough</u> **to determine the type of problem it is, and then set up a structure to capture the information**.
2. **Read the word problem up until a punctuation mark**.
3. **Stop and translate what you just read from words to math.** *Translate phrase by phrase (rather than word by word), just as you would translate a foreign language into English.*
4. **Write it down <u>before</u> you read the next part of the word problem**.
   - **Look for the word "is" which means "equals" and write down the equals sign**.
   - **Tackle the statement in two parts**. Any words that come before the word "is" in a single statement go on the left of the equals sign. Any words that come after the word "is" go on the right of the equals sign. See the next page for a list of words & phrases and what they mean.
   - **Try to write the equation in words first**, **math second**. Most students are better at remembering or figuring out formulas by writing them in words, then placing the values from the question <u>underneath</u> the corresponding words.
   - **Always write down the units**. No "naked numbers."
5. **Now, read up until the next punctuation mark**. Repeat the steps above as needed. A new sentence requires a new equation. *Commas and periods are used for a reason, so use those grammatical pauses to stop and process what you've just read, and translate each phrase into parts of a written formula, constraints on the possible values of a number, etc.*

Off the Charts! Data Interpretation

MATH CLUE WORDS AND WHAT THEY MEAN

Math Operation	Clue Words & Phrases in a Word Problem
Equals	- Is, are, will be. *Any form of the verb "to be" signals the need for an equals sign.* - Equals - Results in - Yields
Greater than >	- ____ is greater than ____ - ____ is more than ____
Greater than or equal to ≥	- ____ is greater than or equal to ____ - ____ is at least ____ - ____ is no less than ____
Less than <	- ____ is less than ____ - ____ is fewer than ____
Less than or equal to ≤	- ____ is less than or equal to ____ - ____ is at most ____ - ____ is no more than ____
Multiply	- Times - By - Of - Product
Divide	- Divided - Split - Per - Out of - Ratio of ____ to ____. *With ratios, set up a fraction. Whichever phrase comes first goes in the numerator. Whichever phrase comes second goes in the denominator.*

Math Operation	Clue Words & Phrases in a Word Problem
Add	- Added to - Plus - More than - Increased by - Sum of
Subtract	- _____ is how much more than _____. *When you see this type of phrasing, you must subtract. The word or phrase that comes first MINUS the word or phrase that comes second.* - Less* - *"17 less a number" means "17 – X"* - Less than* - *"Five less than a number" means "Q - 5"* - Difference of - Difference between - Decreased by

RATIOS & PROPORTIONS

A **ratio** is like a rate: it's a "something per something" which expresses the relationship of two quantities relative to one another.

There are multiple ways to express ratios in words and in math form. All of these phrases are equivalent:

- There are 20 students for every 3 teachers
- The school has a student-teacher ratio of 20:3
- The ratio of students to teachers is 20 to 3
- $\frac{20\ students}{3\ teachers}$

A **proportion** is two ratios set equal to each other. You can use a variable as a placeholder for whichever value is missing, then solve for the value of the variable by cross-multiplying.

Off the Charts! Data Interpretation

TRANSLATING WORDS TO MATH IN RATIO PROBLEMS

A **ratio** is always presented in simplified (reduced) form. The actual value of each of the two numbers could be <u>any multiple</u> of this ratio.

For example, if you know the ratio of boys to girls in a certain class is 3 to 2:

- There are 3 boys for every 2 girls
- The reverse is also true***: There are 2 girls for every 3 boys
- There <u>could be</u> 15 boys and 10 girls (because 15:10 reduces to 3:2)
- There <u>could be</u> 42 boys and 28 girls (because 42:28 reduces to 3:2)
- There <u>could be</u> 66 boys and 44 girls (because 66:44 reduces to 3:2)
- There <u>could be</u> any number of boys and girls such that you have the ratio of 3M:2M, where M is that constant multiple.

***It is critical that you **pay attention to which way you are asked to present a ratio**. The ratio of boys to girls is 3 to 2, but the ratio of girls to boys is the reciprocal, 2 to 3. It's a common test maker trap to offer choices that switch around which part goes where.

- The ratio of A to B means "A:B" or "A over B"
- The ratio of B to A means "B:A" or "B over A"

PART-TO-PART VS. PART-TO-TOTAL RATIOS

Whenever you are presented with a ratio, you will want to understand whether the ratio is a **part-to-part ratio** or a **part-to-total ratio**. Sometimes, you will be able to logically infer from a part-to-part ratio what the part-to-total ratio must be...usually, because there is some **binary condition** (either/or) which logically means that each item is either in one group or the other, but not both. *Progressive sociopolitical values aside, assume anything traditionally accepted as binary, like gender, should be treated as such for the purpose of solving ratio questions on standardized tests.*

For example, if you have: 3 boys for every 2 girls in our class

- This is a **part-to-part ratio** of 3 boys to 2 girls
- You can infer that the whole would equal the total number of boys + girls, so the whole = 5
- Then you have a **part-to-total ratio** of 3 boys to 5 students
- You have another **part-to-total ratio** of 2 girls to 5 students
- Therefore, if you're asked "which of the following could be the number of students in the class," you look for numbers which are multiples of the whole, which is 5. The number of students in the class could be any multiple of 5, such as 15, 30, 65, 185, etc., but cannot be 14 or 22 or 36 because these are not multiples of 5.

For example, if you have: 2 store managers for every 15 store associates

- This is a **part-to-part ratio** of 2 managers to 15 store associates
- Absent any info in the problem to suggest otherwise, you can infer that the whole would equal the total number of store managers + store associates, so the whole = 17
- Then you have a **part-to-total ratio** of 2 managers to 17 store employees
- You have another **part-to-total ratio** of 15 store associates to 17 store employees

Therefore, if you are asked, "which of the following could be the number of store employees," you look for numbers which are multiples of the total, which is 17. The number of store employees could be 34, or 85, or 170, or any other multiple of 17, but cannot be 100 or 75 or anything that is not a multiple of 17.

PERCENTAGES: PERCENT OF AND PERCENT CHANGE

Next, you need to understand the terms **Percent Of** and **Percent Change**.

- This is another fundamental difference you must understand: **The <u>percent of</u> a number is not the same thing as the <u>percent change</u>** (where a percent change could be either some <u>percent more</u> or some <u>percent less</u>). This leads us to a formula:
    - $Percent\ Of = Original\ Amt + (Original\ Amt * Percent\ Change)$
    - Notice here, that the Original Amount shows up in two places. That means this formula can also be re-expressed, by factoring out the Original Amount from each term, as:
    - $Percent\ Of = Original\ Amt * (1 + Percent\ Change)$
    - When you need to solve for the **percent change**, use whichever version of this formula is easier for you to grasp. *Most of the time when students get a question about percent change incorrect, it is because they have put the wrong number in the denominator.*
    - $Percent\ Change = \dfrac{New - Original}{Original}\ or\ \dfrac{Focus - Basis\ of\ Comparison}{Basis\ of\ Comparison}$
- Note, you enter a **percent increase as a positive number** for percent change, and a **percent decrease as a negative number**.

For example, if the regular price of an item is $200 and the sale price of that item is $170, the <u>answers to the following two questions are different percentages</u>.
- The sale price is what percent less than the regular price?
- The regular price is what percent more than the sale price?

The key difference is what goes in the denominator. The most common mistake students make when solving percent change questions is to put the wrong number in the denominator. On multiple choice tests, one of the available-but-wrong answers will almost always be the result of using the wrong denominator.

Back to the example from the preceding page:

- The sale price is what percent less than the regular price?
  - The phrase "what percent less than" means you must find the percent change.
  - The phrase "the sale price" precedes the phrase "what percent less than" so "the sale price" is your focus.
  - The phrase "the regular price" follows the word "than" so "the regular price" is your basis of comparison.
  - $Percent\ Change = \dfrac{Focus\ -\ Basis\ of\ Comparison}{Basis\ of\ Comparison} = \dfrac{Sale\ Price\ -\ Regular\ Price}{Regular\ Price}$
  - $Percent\ Change = \dfrac{\$170\ -\ \$200}{\$200} = \dfrac{-30}{200} = -15\%$
  - A negative percent change means there was a discount or decrease. For your exam, though, you would remove the negative sign and choose a 15% discount or just 15%.

- The regular price is what percent more than the sale price?
  - The phrase "what percent more than" means you must find the percent change
  - The phrase "the regular price" precedes the phrase "what percent more than" so "the regular price" is your focus.
  - The phrase "the sale price" follows the word "than" so "the sale price" is your basis of comparison.
  - $Percent\ Change = \dfrac{Focus\ -\ Basis\ of\ Comparison}{Basis\ of\ Comparison} = \dfrac{Regular\ Price\ -\ Sale\ Price}{Sale\ Price}$
  - $Percent\ Change = \dfrac{\$200\ -\ \$170}{\$170} = \dfrac{30}{170} = \mathbf{17.6\%}$
  - A positive percent change means there was an increase or markup.

PROBABILITY

Term	Definition & Additional Info
Probability	At its simplest, **probability** refers to the likelihood that something occurs. The **probability** of any event happening ranges from: • 0%, or impossible/never --- to --- • 100%, or guaranteed/always The probability of an event happening is the simplified fraction or percentage of the number of **desired outcomes** out of the total number of **possible outcomes**. $$P(Event\ happens) = \frac{\#\ of\ desired\ outcomes}{\#\ of\ possible\ outcomes}$$ The probability of an event happening always equals 100% minus the probability of that event not happening. $$P(Event\ happens) = 1 - P(Event\ doesn't\ happen)$$
Event	What's an **event**? It's some activity of interest. Some examples of common **events** used in probability questions include: • Rolling a 5 on a single fair die (or "number cube") • Selecting a Heart from a standard deck of 52 playing cards • Raining on Tuesday The probability of an event happening can be expressed as either a fraction or a percent. In most cases on the standardized tests, you will want to use the fraction form so that you can take advantage of cancelling.
Odds of *something happening*	Clue words (trigger phrase) that a word problem deals with **Probability**.
Equal chance of *something happening*	Clue words (trigger phrase) that a word problem deals with Probability, and that two or more events have the <u>same probability</u> of occurring.

Term	Definition & Additional Info
When to add vs. multiply probabilities of multiple events	When approaching all **Probability** questions, as well as any **Combinatorics** questions, the word: • **AND means** "multiply" • **OR means** "add" Think about it logically: If two uncertain events need to both "go your way," the <u>likelihood that both do so is lower</u> than if you only needed one of the two events to "go your way." You represent probabilities as either fractions or decimals between 0 and 1, or percentages between 0% and 100%. Anything multiplied by a positive fraction or decimal less than 1 will become smaller.

Independent vs. Dependent Events	Probability questions can involve more than one event. You must determine from the question whether the events are **independent** of one another, or **dependent** on one another.
	
	• **Independent events**: Two events are independent of one another if one event happening does not influence or impact the likelihood that the other. They're unrelated. For example:
	  ○ The probability that you draw a card that's 7 or lower from a standard deck of 52 cards, and the probability that you roll a total of 8 on a pair of dice (number cubes).
	
	• **Dependent events**: Two events are dependent on one another if one event happening makes another event more or less likely to happen. They're related. For example:
	  ○ The probability that it is raining outside, and the probability that you bring an umbrella to work.

Term	Definition & Additional Info

Mutually Exclusive vs. Not Mutually Exclusive Events	Probability questions can involve more than one event. You must determine from the question whether the events are **mutually exclusive** of one another, or **not mutually exclusive**.
	
	- **Mutually exclusive events**: Two events are mutually exclusive of one another if one event happens, then the other event definitely does not happen. The two events cannot occur at the same time. For example:
	  - Raining vs. not raining
	  - Happy vs. not happy
	
	- **Not mutually exclusive events**: Two events are not mutually exclusive of one another if it is possible that both events can happen at the same time. For example:
	  - Wearing shorts and wearing sandals. You can wear shorts with running shoes but you can also wear them with sandals. So, these events are not mutually exclusive.
	  - Going to grad school and working full-time. You can obviously do one, both, or neither at the same time. So, these events are also not mutually exclusive.

Term	Definition & Additional Info
Probability of Multiple Events	For any situation with two events, A and B, you have multiple scenarios to evaluate: • Neither Event A nor Event B happens (Neither) • Event A happens, but Event B does not happen (A only) • Event B happens, but Event A does not happen (B only) • Both Event A and Event B happen (A and B) When you have more than one event, you can determine the **probability that <u>at least 1 of the 2</u> events happens**, which means either-or-both, using the following formula. You subtract off the probability of both A and B happening because it will otherwise be double-counted. *This potential double-counting is similar to what you may have seen in Venn Diagram problems; in both situations, simply adding the instances of both events separately would cause you to double-count the overlap.* $$P(at\ least\ 1\ of\ the\ events) = P(A) + P(B) - P(A\ and\ B)$$ You could restate this in plain language as: $$P(at\ least\ 1\ of\ the\ events) = P(A) + P(B) - P(both)$$ You can also determine the **probability that either A or B happens, but not both**, using the following formula. In other words, this is the **probability that <u>exactly one</u> of the events happens**. You subtract off two times the probability of both A and B happening because it is double-counted, and you want to fully exclude it. $$P(either\ A\ or\ B) = P(A) + P(B) - 2*P(A\ and\ B)$$ You could restate this in plain language as: $$P(either) = P(A) + P(B) - 2*P(both)$$ If events A and B are **mutually exclusive**, then: • The last term, P(both) will equal zero If events A and B are **not mutually exclusive**, then: • The last term, P(both) will not equal zero • You'll need to find P(both) by multiplying the individual probabilities of events A and B. $$P(both) = P(A)*P(B)$$

THINKING THROUGH LOGIC: WORDS FIRST, MATH SECOND

The basic question, "**What is the probability of**...*a certain situation(s) with various parameters or constraints*?" are the clue words which should trigger you to think in terms of Probability.

- The activity or context of the question will change – flipping one or more coins; rolling one or more "number cubes" (aka, dice); drawing one or more cards from a standard 52-card deck; rain or no rain; drawing a marble of a certain color from a bag of several marbles; drawing names from a raffle; selecting the Powerball jackpot numbers; the odds of winning certain casino games; and others.

- The parameters or constraints will change – and you must pay strict attention to these details.
    - Are you making **one or more than one selection**?
    - If there are **multiple events**, does outcome of one event have an impact or no impact on the probability of the second event? In other words, are the two events **independent** or **dependent**?
    - Is **replacement** allowed or not?
        - The best example to understand **replacement** vs. **no replacement**: A raffle with sequential drawings for several prizes.
        - If your name is drawn for the first prize, will your name "go back in the hat" to be eligible for any of the other, often bigger/better, prizes, or will you become ineligible for the subsequent drawings?
    - If the events are **sequential**, whether the ability to be chosen on the 3rd selection is predicated upon not being chosen for the 1st or 2nd selection.
        - For example, if there are 3 student council positions of President, Vice President, and Secretary, and the 3 positions are elected in sequential order, the probability that a certain person is selected for the role of Secretary means that to become Secretary, this person:
            - Must not be selected for the role of President, and
            - Must also not be selected for the role of VP, and
            - Is selected for the role of Secretary
            - Thus, the logic in **Words First**, **Math Second** becomes:

                $$P(\text{Elected Sec}) = P(\text{Not Pres}) * P(\text{Not VP}) * P(\text{Yes Sec})$$

        - *The example above is really another way of saying "there is no replacement."*

- When you encounter more complex scenarios, try to articulate the scenario(s) which fit the question in your own words. You will want to pay attention to where you include the words which trigger math operations:
    - **EITHER/OR** = Add, often the probability of two cases which fit the described scenario
    - **AND** = Multiply, often that this-AND-that need to happen in scenario.

You should use the **Words First**, **Math Second approach** for structuring your approach.

Some of the most common scenarios described in word problems testing your understanding of Probability include:

- Flipping one or more coins
- Rolling one or more number cubes (i.e., a single die or ≥ 2 dice)
- Drawing colored marbles from a jar
- Drawing names from a hat

Selecting a certain color of shirt or pants from a closet or drawer

STATISTICS & WEIGHTED AVERAGES

If you're preparing for one of the standardized tests, **statistics** is probably one of the topics which seems more daunting than it really is. That's because these topics are often glossed over in high school math classes, and not all college majors require a course in statistics.

Fortunately, there is a fairly limited scope of statistics concepts tested on the major college and graduate admissions standardized tests. You will need to be familiar with how to find, utilize, and compare:

- The **measures of centrality**: Finding and using the mean, median, and mode.
- The **measures of dispersion**: Finding and using the range, quartile range, and standard deviation.

$$Range\ of\ a\ Set = Maximum\ Value - Minimum\ Value$$

- The **normal distribution** (bell curve) and its components: mean, standard deviation, variance, and percentile.
- Other measures of **relative standing/ranking**: Decile, quintile, and quartile.

UNDERSTANDING THE MEASURES OF CENTRALITY

Here's what you need to know about the three measures of centrality used to describe a data set (or a population):

	Mean	Median	Mode
Definition	The **arithmetic average** of a data set	The **middle value**, when the data set is listed in order from smallest to largest	The term which **appears most frequently** in the data set
How to find this measure	= (Sum of terms) / (# of terms)	• First, list them in order by size • If the set has an odd number of terms, it's the term in the middle. • If the set has an even number of terms, take the average of the two terms in the middle.	If the set is provided in a list like this: {a, b, c, c, d, c, e} Then it's the term which appears most frequently. If the data is presented in a histogram, look for the tallest column.
Practical usage	The **mean** is the most commonly used of the three measures of centrality. It is most useful and relevant when the data set is reasonably symmetric, and/or there is both a minimum and a maximum possible value for the data. You're probably quite familiar with means (arithmetic averages): • Your GPA • Average test scores (range of 0 to 100) for your classes • Average height	The **median** is most useful and relevant when the data set is skewed, and there are a few outliers (extremes) which cause the "average" to be overstated. **Household income** (**HHI**) is the most common stat that uses median, instead of an average. Why? If you include the income of billionaire CEOs, the "average" HHI of your state is very high. The median is more relevant for planning things like public transportation.	The **mode** is used much less often than the mean or median. It's more relevant for things like consumer surveys, where you're looking for frequency of an answer more than an exact value (or where the answers are split into two camps).

WEIGHTED AVERAGES

A **weighted average** is an overall average across two or more groups. For example, if a large middle school has three 7th grade math teachers:

- Each teacher will have his or her own class average on the final exam.
- The average of all 7th grade math students is **probably not** equal to the simple average of the three class averages. Why? Because the classes may not be equally sized. One class may have a few more students than the other two classes. Let's call these classes Groups A, B, and C.
- There are **two variations of the weighted average formula**. You can use either the **number of observations** or the **percent of observations** in each group.

$$Wtd.Avg = \frac{(Group\ A\ avg\ *\ \#\ in\ A) + (Group\ B\ avg\ *\ \#\ in\ B) + (Group\ C\ avg\ *\ \#\ in\ C)}{\#\ in\ A + \#\ in\ B + \#\ in\ C}$$

OR

$$Wtd.Avg = (Group\ A\ avg\ *\ \%\ in\ A) + (Group\ B\ avg\ *\ \%\ in\ B) + (Group\ C\ avg\ *\ \%\ in\ C)$$

- From the first formula, you can see that to find the weighted average – or the average of all 7th graders – you need to know each of the three class averages and the number of students in each class.
- Using the data below, we'll find the average score for all 7th grade math students.

Teacher	Class Average	# of Students
Ms. Anderson	75	30
Mr. Baker	88	20
Ms. Cook	84	25

$$Weighted\ Avg = \frac{(75 * 30) + (88 * 20) + (84 * 25)}{30 + 20 + 25}$$

$$Weighted\ Avg = \frac{2250 + 1760 + 2100}{75} = \frac{6110}{75} = 81.46$$

- Some facts to remember about weighted averages, which are useful **heuristics** (or rules of thumb) to help you efficiently eliminate tempting-but-wrong answer choices:
    - The weighted average cannot be less than the smallest group average
    - The weighted average cannot be more than the largest group average
    - The weighted average will only equal the simple average of the group averages if the groups are equally sized. Otherwise, the weighted average will not equal the average of the group averages.

Off the Charts! Data Interpretation

PERCENTILES, DECILES, QUINTILES, AND QUARTILES

In a data set with any kind of distribution (normal or otherwise), you use one of three terms to express a person or observation's **relative standing** compared to all others in the same set.

You might already be familiar with **percentiles** from your past experience with standardized tests – you'll get a numeric score and a percentile ranking.

- **Percentile**: The percentile associated with a particular score or data point expresses what percent of the sample population scored the same or less than that person or observation. This could be restated as the person scored the same or better than what percent of the sample population. You can multiply your percentile rank times the size of the sample population to get an exact value for the number of students whom you beat.
    - For example, on a recent administration of the ACT, a composite score of 26 would give you an 83rd percentile rank. That means, this student who scored a 26 performed the same as or better than 83% of all students taking that test on the same date.
    - If 50,000 students took the ACT on the same date, then this student scored the same as or better than 83% * (50,000) = 41,500 other students.
    - *Note: Absolute scores on standardized tests can fluctuate up or down over time, or across different exam dates, so the percentile ranking is what gives meaning to your absolute score.*

Sometimes, you would prefer to cluster or group the sample population into equally sized units. For that purpose, you use Decile, Quintile, and/or Quartile.

In the real world, you may see deciles used to describe various types of population or market segments.

- *Decile: Using the Latin roots, you know that calculating deciles means splitting the population into 10 groups, when ordered from least to greatest. So, each decile contains (100%/10) = 10% of the population.*
    - *The 1st decile would represent the lowest score up to the 10th percentile;*
    - *The 2nd decile would represent the 11th percentile up to the 20th percentile;*
    - *The 3rd decile would represent the 21st percentile up to the 30th percentile;*
    - *And so on…*

In the real world, you may see quintiles used to describe household income data and its association to socioeconomic classes, with such terms as "Working Class", "Lower Middle Class", "Middle Class", "Upper Middle Class" and "Affluent" (or other similar terms).

- *Quintile: Using the Latin roots, you know that calculating quintiles means splitting the population into 5 groups, when ordered from least to greatest. So, each quintile contains (100%/5) = 20% of the population.*
    - *The 1st quintile would represent the lowest score up to the 20th percentile;*

- The 2nd quintile would represent the 21st percentile up to the 40th percentile;
- The 3rd quintile would represent the 41st percentile up to the 60th percentile;
- And so on…

In the real world, you may see quartiles used to provide information about the range of the middle of the data set, after eliminating the upper and lower ends of the set. For example, many colleges present their 25th and 75th percentile test scores and GPAs for admitted students (also called the "middle 50 percent", which is found by finding the scores at the bottom of the 2nd quartile and the top of the 3rd quartile) to help students understand whether their application is competitive for their desired school. For example, if a school had an average ACT score of 31.3, some students with a 29 or 30 might be too intimidated to apply, so if the same school shares a middle 50 percent range of scores from 27 to 33, those students with a 29 or 30 would better understand that they're right in the sweet spot of that school's applicant pool.

- Quartile: Using the Latin roots, you know that calculating quartiles means splitting the population into 4 groups, when ordered from least to greatest. So, each quartile contains 25% of the population.
  - The 1st quartile would represent the lowest score up to the 25th percentile
  - The 2nd quartile would represent the 26th percentile up to the 50th percentile
  - The 3rd quartile would represent the 51st percentile up to the 75th percentile
  - The 4th quartile would represent the 76th percentile up to the 100th percentile

Off the Charts! Data Interpretation

CHAPTER 13 CLOSING

Congratulations! You have finished the entire book and have learned quite a few new approaches for solving complex data interpretation questions along the way.

The next task at hand is to practice applying these approaches to a sufficient number of problems for your specific standardized test. As you gain experience applying the strategies in **Off the Charts! Data Interpretation** to solve different questions about the many varieties of tables, charts, and graphs, you will gain a tremendous amount of confidence in yourself and your data interpretation abilities!

The hardest problems you'll see on your test are those which combine multiple topics. Just remember, the difficulty comes in recognizing which topics are in play, outlining an approach, and working through a few more steps than usual. You may find, as many of my former students have, that this book has changed your relationship with math and data interpretation for the better and changed the way you think.

If you'd like to reach out to this book's author to share your success story or to engage her for additional one-on-one assistance, contact her at:

coachinginquiry@sparkadeptation.com

www.ingramcontent.com/pod-product-compliance
Lightning Source LLC
Chambersburg PA
CBHW061124070526
44584CB00033B/4215